PROBLEMS AND SOLUTIONS
IN
ELECTRICAL MACHINE

PROFESSOR SATISH DIXIT

DEPARTMENT OF ELECTRICAL
ENGINEERING
ICFAI UNIVERSITY RAIPUR
INDIA

CONTENTS

Emf Generated

Where E_g = generated emf, P =No of poles ,Z = Total no of conductors, N= Speed in revolution per minute, A = No. of parallel path (For wave wound A=2, And for lap wound A=P)

Separately Excited

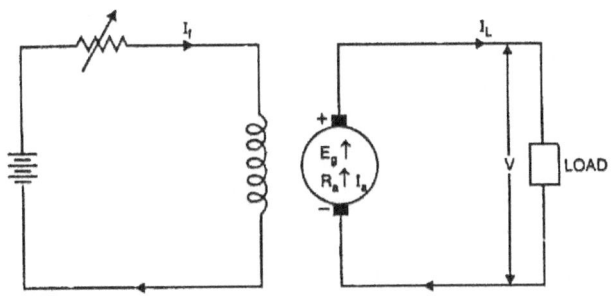

Fig.1

Armature current, $I_a = I_L$
Terminal voltage, $V = E_g - I_aR_a$
Electric power developed $= E_gI_a$
Power delivered to load $= E_gI_a - I^2{}_aR_a = VI$

Dc Series generator

It has a few turns of thick wire having low resistance as shown in (Fig.2)

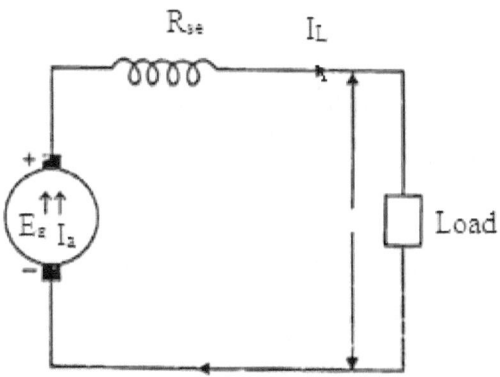

Fig.2

Armature current, $I_a = I_{se} = I_L = I(\text{say})$

Terminal voltage, $V = E_g - I(R_a + R_{se})$

Power developed in armature $= E_g I_a$

Power delivered to load$= E_g I_a - I^2_a (R_a + R_{se}) = V I_a$

DC Shunt Motor

The shunt field winding has many turns of fine wire having high resistance as shown in (Fig.3).
Shunt field current, $I_{sh} = V/R_{sh}$
Armature current, $I_a = I_L + I_{sh}$

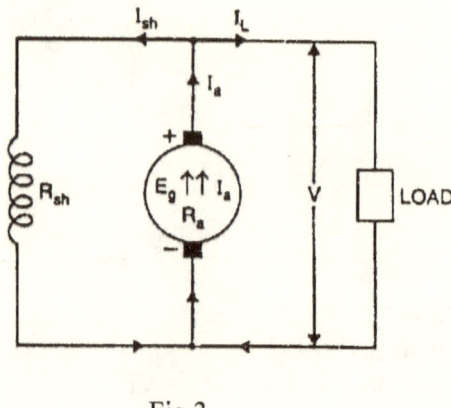

Fig.3

Terminal voltage, $V = E_g - I_a R_a$

Power developed in armature $= E_g I_a$

Power delivered to load $= V I_L$

Compound generator

Short Shunt in which only shunt field winding is in parallel with the armature winding(Fig.4)

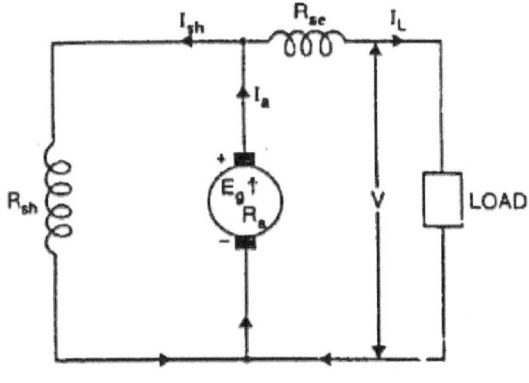

Fig.4

Series field current, $I_{se} = I_L$

Shunt field current, $I_{sh} = (V + I_{se}R_{se})/R_{sh}$
Terminal voltage, $V = E_g - I_aR_a - I_{se}R_{se}$
Power developed in armature $= E_gI_a$
Power delivered to load $= VI_L$

Long shunt

Long Shunt in which shunt field winding is in parallel with both series field and armature winding As shown in (Fig5)
Series field current, $I_{se} = I_a = I_L + I_{sh}$
Shunt field current, $I_{sh} = V/R_{sh}$

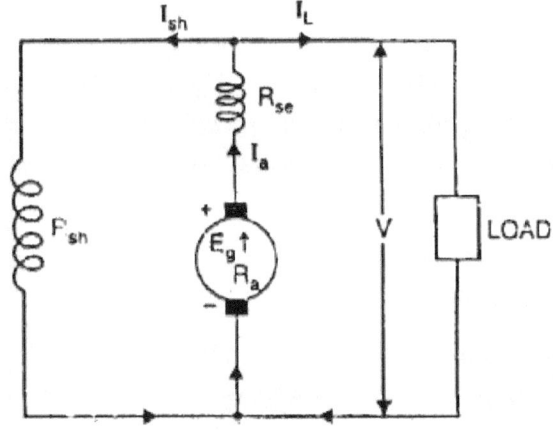

Fig.5

Terminal voltage, $V = E_g - I_a(R_a + R_{se})$
Power developed in armature $= E_g I_a$
Power delivered to load $= VI_L$

Losses in a D.C. Machine

These losses occur due to currents in the various windings of the machine.

    (i)   Armature copper loss $= I^2_a R_a$
    (ii)   Series field copper loss $= I^2_{se} R_{se}$
    (iii)   Shunt field copper loss $= I^2_{sh} R_{sh}$

Iron or Core losses

These losses occur in the armature of a d.c. machine and are due to the rotation of armature

Hysteresis loss

While machine is rotating under magnetic field armature is subjected to magnetic field reversals as it passes under successive poles.

Hysteresis loss, $P_h = B^{1.6}_{max} f V$ watts

To reduce this loss in a d.c. machine, armature core is made of materials which have a low value of Steinmetz hysteresis co-efficient e.g., silicon steel.

Eddy current loss

When machine starts rotating then voltage not only induced in armature but there are also voltages induced in the armature core. These voltages produce circulating currents in the armature core These are called eddy currents and power loss due to their flow is called eddy current loss.

$P_e = B^2_{max} f^2 t^2 V$ watts

Mechanical losses

These losses are due to friction and windage.
(i) friction loss, (bearing friction, brush friction) etc.

(ii) windage loss, (air friction of rotating armature).

(iii)Iron losses and mechanical losses together are called stray losses.

Mechanical efficiency

$\eta_m = E_g I_a$ / Mechanical power input

Electrical efficiency

$\eta_\varepsilon = V I_L / E_g I_a$

Commercial or overall efficiency

$\eta_c = V I_L$ / Mechanical power input

$\eta_c = \eta_\varepsilon \times \eta_m$

Condition for Maximum Efficiency

Variable loss = Constant loss

$I^2_L R_a = W_c$

D.C. Generator Characteristics

Open Circuit Characteristic (O.C.C.)

The relation between the generated e.m.f. at no-load (E_0) and the field current (I_f) at constant
speed. It is also known as magnetic characteristic or no-load saturation curve(Fig.6).

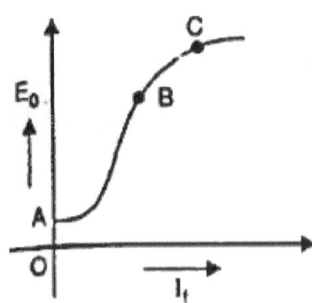

Fig.6

(i) When the field current is zero, there is some generated e.m.f. OA. This is due to the residual

magnetism

(ii) upto point B in the curve, the curve is linear. It is because in this range, reluctance of iron is

negligible as compared with that of air gap. The air gap reluctance is constant and hence

linear relationship.

(iii) After point B on the curve, the reluctance of iron also start working. It is because at higher

flux densities, μ_r for iron decreases and reluctance of iron is no longer negligible.

(iv)After point C on the curve, the magnetic saturation of poles begins and E_0 tends to level off.

(v) The maximum field circuit resistance (for a given speed) with which the shunt generator would just excite is known as its critical field resistance.

Characteristics of Series Generator;-Shown in (Fig7)

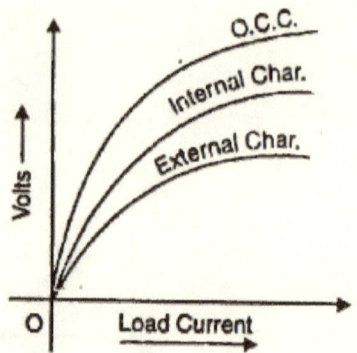

Fig.7

Internal characteristic generated e.m.f. E on load and armature current.

External characteristic relation between terminal voltage and load current I_L.

The internal and external characteristics of a d.c. series generator can be plotted from one another as shown in Fig.8

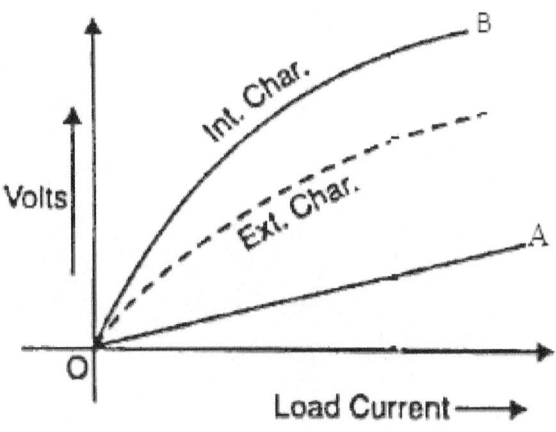

Fig.8

In the above figure, line OA represent the resistance of the whole machine i.e. $R_a + R_{se}$.

Characteristics of a Shunt Generator(Fig 9)

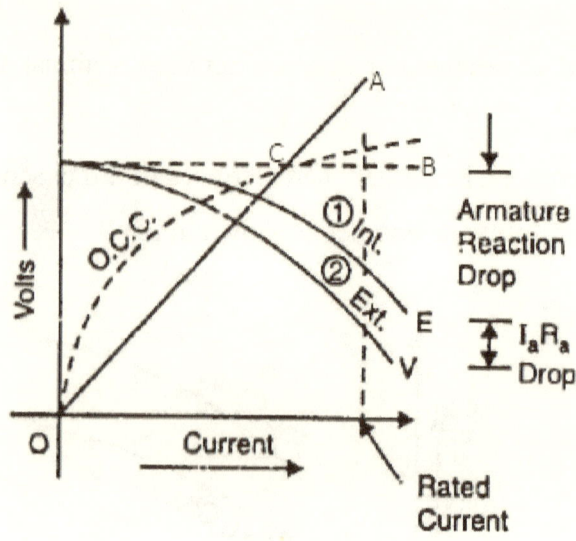

Fig. 9

Compound Generator Characteristics (Fig10)

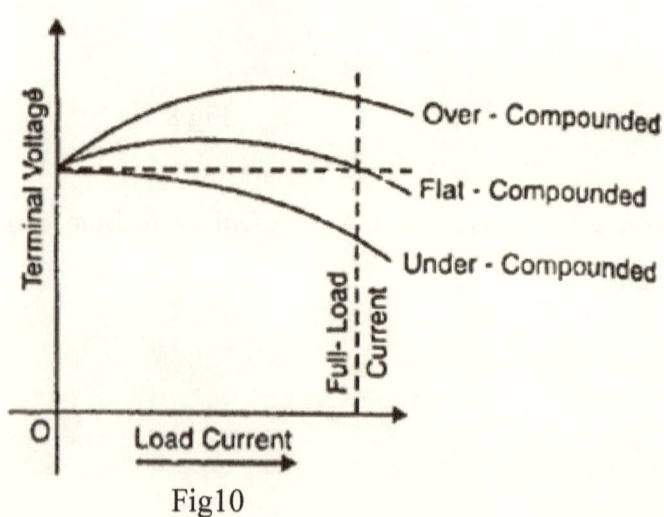

Fig10

%Voltage regulation = V_{NL} - V_{FL} / V_{FL} X 100

V_{NL} = Terminal voltage of generator at no load

V_{FL} = Terminal voltage of generator at full load

D.C. Motor Principle

The direction of this force is given by Fleming's left hand rule and magnitude is given by;

F = BIl newtons

Power Equation = $VI_a = E_bI_a + I^2_aR_a$

Condition For Maximum Power in Dc Machine

$E_b = V/2$

Shunt Motor:- R_{sh} is a shunt field resistance (Fig 11).

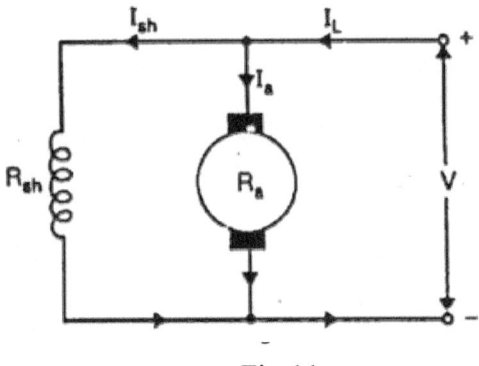

Fig.11

Series motor:- R_{se} series field resistance(Fig.12)

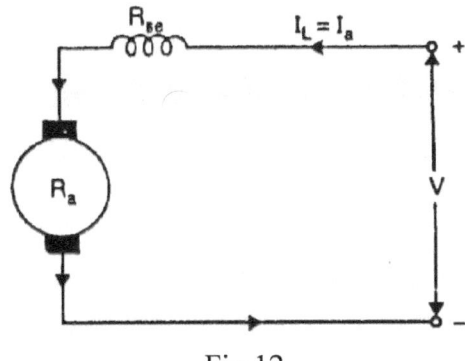

Fig.12

Armature Torque of D.C. Motor

$T_a = 0.159\ P\phi Z(\ I_a/N)$ N-m

Or $T_a = 9.55\ E_b(\ I_a/N)$ N-m

Shaft Torque (Tsh)

Ta - Tsh = 9.55 X Iron and frictional losses/N

Speed of a D.C. Motor

$$\frac{N_2}{N_1} = \frac{E_{b2}}{E_{b1}} \times \frac{\phi_1}{\phi_2}$$

For a shunt motor, flux practically remains constant so that $\phi_1 = \phi_2$.

$$\frac{N_2}{N_1} = \frac{E_{b2}}{E_{b1}}$$

For a series motor, ϕ is directly propotional to the I_a.

$$\frac{N_2}{N_1} = \frac{E_{b2}}{E_{b1}} \times \frac{I_{a1}}{I_{a2}}$$

Speed Regulation

% Speed regulation $= \dfrac{N_0 - N}{N} \times 100$

where N_0 = No - load .speed

N = Full - load speed

Shunt Motor

Torque – Armature current characteristic (Fig13)

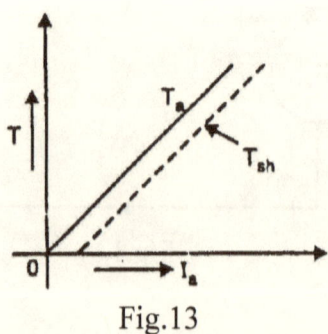

Fig.13

Speed – Torque characteristics (Fig.14)

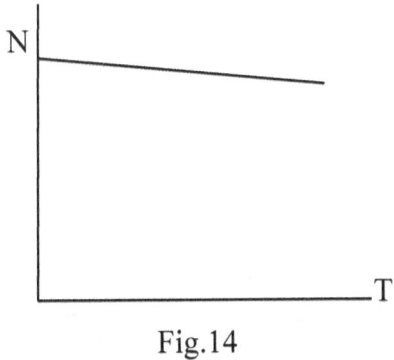

Fig.14

Series Motor

Torque – Armature current characteristic(Fig15).

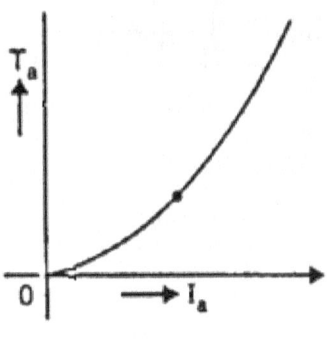

Fig.15

Speed – Torque characteristics in figure 16 given below

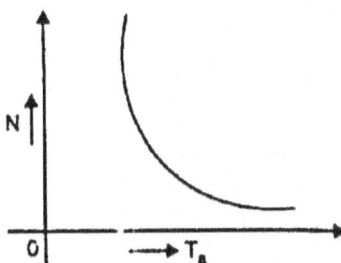

Mechanical efficiency

η_m = Output power of motor/ $E_b I_a$

Electrical efficiency

$\eta_\varepsilon = E_b I_a / V I_L$

Commercial or overall efficiency

η_c = Output power of motor / $V I_L$

$\eta_c = \eta_\varepsilon \times \eta_m$

Ideal Transformer
An ideal transformer is one that has

(i) no winding resistance

(ii) no leakage flux i.e., the same flux links both the windings

(iii) no iron losses (i.e., eddy current and hysteresis losses) .

E.M.F. Equation of a Transformer

$E_2 = 4.44 \, f \, N_2 \, \phi_m$
$E_1 = 4.44 \, f \, N_1 \, \phi_m$

Voltage Transformation Ratio (K)

$$\frac{N_2}{N_1} = \frac{E_2}{E_1} = \frac{V_2}{V_1} = \frac{I_1}{I_2} = K$$

Resistance and Inductance of secondary referred to primary (Fig.17)

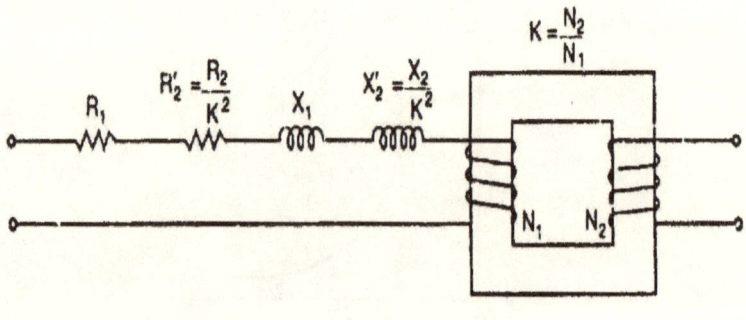

Fig.17

$R_{01} = R_1 + R'_2$

$X_{01} = X_1 + X'_2$

Resistance and Inductance of primary Referred to secondary (Fig.18)

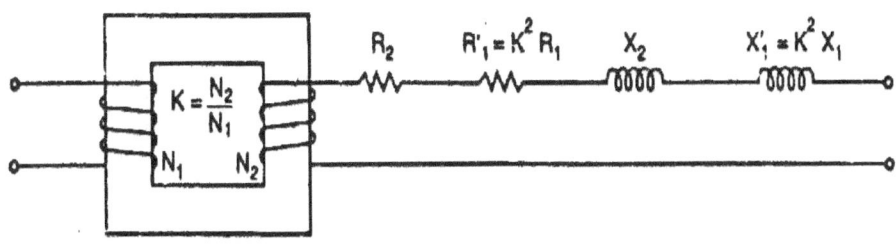

Fig.18

$R_{02} = R_2 + R'_1$

$X_{02} = X_2 + X'_1$

Open-Circuit or No-Load Test

Performed on low voltage side, High voltage side remain open circuit. Rated voltage of low voltage side is applied and 2% to 5 % of rated current is applied.

Iron losses, Pi = Wattmeter reading = W_0

No load current = Ammeter reading = I_0

Applied voltage = Voltmeter reading = V_1

Input power, $W_0 = V_1 I_0 \cos \phi_0$

Short-Circuit or Impedance Test

Performed on high voltage side, low voltage side remain short circuit. Rated current of high voltage side is applied and 5% to 12 % of rated voltage is applied.

Full load Cu loss, P_C = Wattmeter reading = W

Applied voltage = Voltmeter reading = V_{SC}

F.L. primary current = Ammeter reading = I_1(If test performed on primary side)

$$Z = \frac{V_{sc}}{I_1}$$

$$\cos\phi_1 = \frac{P_c}{V_{sc}I_1}$$

Hysteresis loss, $P_h = K_h B1.6_{max} f$ watts, eddy current loss. $P_e = K_e B^2_{max} f^2 t^2$ watts

Efficiency $= \dfrac{output\ power}{input\ power}$

Efficiency$(\eta)\%$ $= \dfrac{x \times KVA \cos\phi \times 1000}{x \times KVA \cos\phi \times 1000 + P_i + x^2 P_c} \times 100$

x is known as fraction of load.

Condition for Maximum Efficiency

Iron losses = Copper losses

$P_c x^2 = P_i$

Efficiency $= \dfrac{KWh\ output\ in\ 24\ hour}{KWh\ input\ in\ 24\ hour}$

Autotransformer

Power transferred inductively = Input X (1 - K)

Power transferred conductively = Input - Input (1 - K)

= Input [1 - (1 - K)]

= K X Input

Weight of Cu in autotransformer (W_a)

= (1 - K) X Weight in ordinary transformer (W_o)

(a)An autotransformer requires less Cu than a two-winding transformer of similar rating.

(b)An autotransformer operates at a higher efficiency than a two-winding transformer of similar rating.

(c) An autotransformer has better voltage regulation than a two-winding transformer of the same rating.

The Speed of Rotation of a Synchronous Generator

$$N_s = \frac{120f}{p}$$

The sum of leakage reactance and armature reaction reactance is called synchronous reactance Xs. Under this condition impedance of the armature winding is called the synchronous impedance Zs.

synchronous reactance $X_s = X_1 + X_a$ Ω per phase
and synchronous impedance $Z_s = R_a + j X_s$ Ω per phase

(i)Unity power factor load (Fig.19)

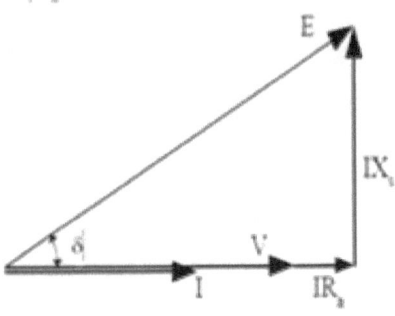

Fig.19

Under unity power factor load: $E_{ph} = (V + IR_a) + j (IX_S)$

$$E_{ph} = \sqrt{(V + IR_a)^2 + (IX_s)^2}$$

(ii)Zero power factor lagging (Fig.20)

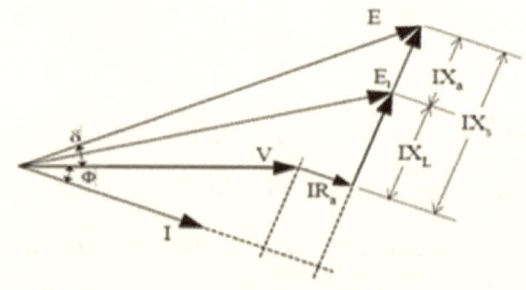

Fig.20

$$E_{ph} = \sqrt{(V\cos\phi + IR_a)^2 + (V\sin\phi + IX_s)^2}$$

(iii)Zero power factor leading (Fig.22)

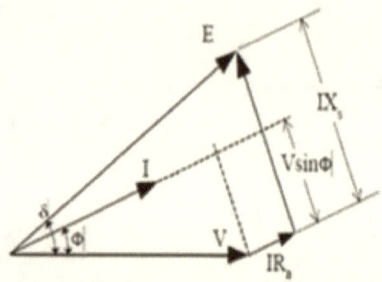

Fig.21

$$E_{ph} = \sqrt{(V\cos\phi + IR_a)^2 + (V\sin\phi - IX_s)^2}$$

Open Circuit Characteristic (O.C.C.) & Short Circuit Characteristic (S.C.C.) (Fig.22)

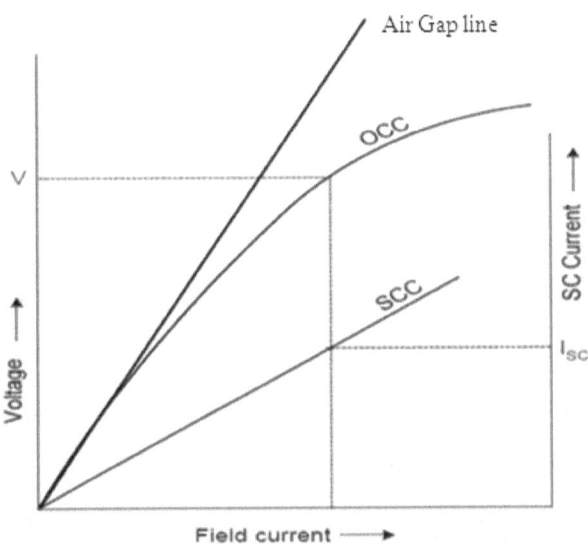

Fig.22

Short-Circuit Ratio:

The short-circuit ratio is defined as the ratio of the field current required to produce rated volts on open circuit to field current required to circulate full-load current with the armature short-circuited. Short-circuit ratio = I_{f1}/I_{f2}.

synchronous impedance Z_s:

The ratio of open circuit voltage to the short circuit current at a particular field current, or at a field current responsible for circulating the rated current is called the synchronous impedance

Hence $Z_s = (V_{oc}) / (I_{sc})$ for same I_f

Direct axis and quadrature axis Reactance

In case salient pole machines the air gap is non uniform and it is smaller along pole axis and is larger along the inter polar axis. These axes are called direct axis or d-axis and quadrature axis or q-axis. Hence the effect of mmf when acting along direct axis will be different than that when it is acting along quadrature axis. Hence the reactance of the stator cannot be same when the mmf is acting along d - axis and q- axis

Synchronous Motor:- sometimes called synchronous condenser.

synchronous condenser:- An over excited synchronous motor operates at unity or leading power factor. Generally, in large industrial plants the load power factor will be lagging. The specially designed synchronous motor running at zero load, taking leading current, approximately equal to 90°. When it is connected in parallel with inductive loads to improve power factor, it is known as synchronous condenser.

Hunting:

Sudden changes of load on synchronous motors may sometimes set up oscillations that are superimposed upon the normal rotation, resulting in periodic variations of a very low frequency in speed. This effect is known as hunting or phase-swinging.

Damper winding:

The tendency of hunting can be minimized by the use of a damper winding. Damper windings are placed in the pole faces. No emfs are induced in the damper bars and no current flows in the damper winding, which is not operative

Torque developed in Motor:

Mechanical power is given by $P_m = 2\pi N_s T_g/60$. where N_s is the synchronous speed and the T_g is the gross torque developed.

Shaft output torque $T_{sh} = 60 \times P_{out}/2\pi N_s$

$T_{sh} = 9.55\ P_{out}/N_s$ N-m

V curve of synchronous motor

Graphs of armature current vs. field current of synchronous motors are called V curves. It can be easily noted from these curves that an increase in shaft loads require an increase in field
excitation in order to maintain the power factor at unity. The lagging power factor operation is electrically equivalent to an inductor and the leading power factor operation is electrically equivalent to a capacitor.(Fig.23)

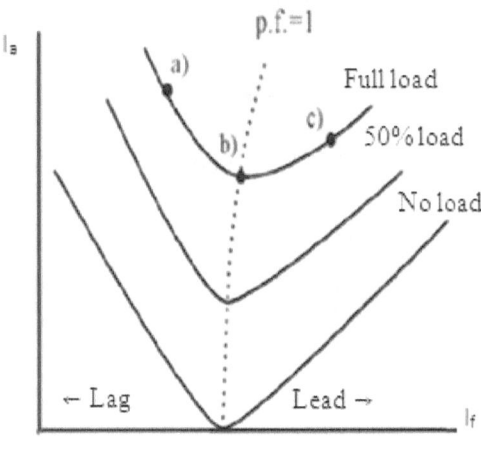

Fig.23

Inverted V curve of synchronous motor

Plots of power factor vs. field current of synchronous motors are called inverted V curves and are shown in Figure below for different values of synchronous motor loads.(Fig.24)

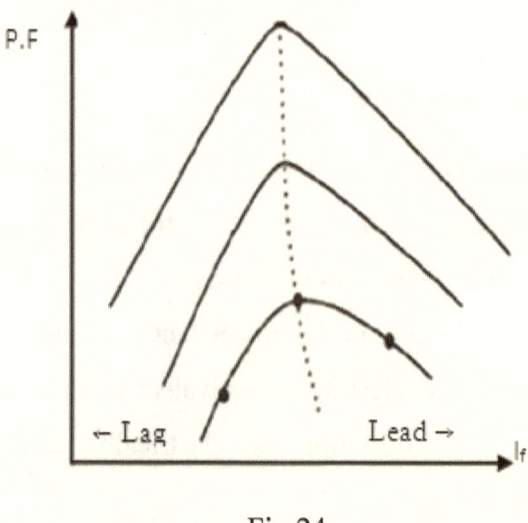

Fig.24

Induction Motor Concepts

The Development of Induced Torque in an Induction Motor

When current flows in the stator, it will produce a magnetic field in stator as such that $\mathbf{B_s}$ (stator magnetic field) will rotate at a speed:

$$n_{sync} = \frac{120 f_e}{P}$$

Where f_e is the system frequency in hertz and P is the number of poles in the machine. This rotating magnetic field $\mathbf{B_s}$ passes over the rotor bars and induces a voltage in them. The voltage induced in the rotor is given by:

$e = (\mathbf{v} \times \mathbf{B}) \, l$

An induction motor can never reach synchronous speed. So sometimes it is called **Asynchronous motor**.

The Concept of Rotor Slip

The induced voltage at the rotor bar is dependent upon the relative speed between the stator magnetic field and the rotor. This can be easily termed as slip speed:

$n_{slip} = n_{sync} - n_m$

Where n_{slip} = slip speed of the machine
       n_{sync} = speed of the magnetic field.
       n_m = mechanical shaft speed of the motor.

Slip may also be described in terms of angular velocity, ω.

$$s = \frac{\omega_{sync} - \omega_m}{\omega_{sync}} \, x100\%$$

% age slip s = $\dfrac{N_s - N}{N_s} \times 100$

the rotor speed: $N_m = (1-s)N_s$
                     $\omega_m = (1-s)\omega_s$

Rotor Current Frequency

f =sf

Power and Torque in an Induction Motor

The input current to a phase of the motor is:

$$I_1 = \frac{V_\phi}{Z_{eq}}$$

Where
$$Z_{eq} = R_1 + jX_1 + \cfrac{1}{G_C - jB_M + \cfrac{1}{\cfrac{V_2}{s} + jX_2}}$$

The stator copper losses, the core losses, and the rotor copper losses can be found.

The stator copper losses in the 3 phases are: $P_{sc} = 3\ I_1{}^2\ R_1$

The core losses : $P_{Core} = 3\ E_1{}^2\ G_C$

So, the air gap power: $P_{AG} = P_{in} - P_{SC} - P_{core}$

The air-gap power can be consumed is in the resistor R_2/s. Thus, the air-gap power:

$$P_{AG} = 3I_2{}^2 \frac{R_2}{s}$$

The actual resistive losses in the rotor circuit are given by:

$P_{RC} = 3\ I_R{}^2\ R_R$

Since power is unchanged when referred across an ideal transformer, the rotor copper losses can also be expressed as:

$P_{RC} = 3\ I_2{}^2\ R_2$

After stator copper losses, core losses and rotor copper losses are subtracted from the input power to the motor, the remaining power is converted from electrical to mechanical form. The power converted, which is called developed mechanical power is given as:

$$P_{conv} = P_{AG} - P_{RCL}$$

$$= 3I_2^2 \frac{R_2}{s} - 3I_2^2 R_2$$

$$= 3I_2^2 R_2 \left(\frac{1}{s} - 1\right)$$

$$P_{conv} = 3I_2^2 R_2 \left(\frac{1-s}{s}\right)$$

the lower the slip of the motor, the lower the rotor losses. Also, if the rotor is not turning, the slip is s=1 and the air gap power is entirely consumed in the rotor.

$$P_{conv} = P_{AG} - P_{RCL}$$
$$= P_{AG} - sP_{AG}$$
$$P_{conv} = (1\text{-}s)\, P_{AG}$$

Finally, if the friction and windage losses and the stray losses are known, the output power:

$$P_{out} = P_{conv} - P_{FW} - P_{other}$$

The induced torque in a machine was defined as the torque generated by the internal electric to mechanical power conversion. This torque differs from the torque actually available at the terminals of the motor by an amount equal to the friction and windage torques in the machine. Hence, the developed torque is:

$$\tau_{ind} = \frac{P_{conv}}{\omega_m}$$

Separating the Rotor Copper Losses and the Power Converted in an Induction Motor's Equivalent Circuit Since Air Gap power would require R_2/s and rotor copper loss require R_2 element. The difference between the air gap power and the rotor copper loss would give the converted power, hence;

$$R_{conv} = \frac{R_2}{s} - R_2 = R_2 \left(\frac{1-s}{s} \right)$$

Qu(1) What is a Transformer?

ANS: A transformer is an electrostatic device which transfer electrical energy from one circuit to another without changing the frequency and it works on the principle of electromagnetic induction'

Qu(2) Is it Possible to Change Three Phase to Two Phase or Vice-Versa with Standard Transformers?

ANS: Yes. This is a very practical application it is done by using scott 240 volts three phase to 240 volts two phase. Please refer to us for an exact schematic.

Qu(3) How Does a Transformer Work?

ANS: A transformer works on the magnetic induction principle. It has no moving parts and is a completely static state device. It consists, in its simplest form, of two or more coils of insulated wire wound on a laminated steel core. When voltage is introduced to one coil, called the primary, it magnetizes the iron core. A voltage is induced in the other coil, called the secondary or output coil. The change of voltage (or voltage ratio) between the primary and secondary depends on the turns ratio of the two coils.

Qu(4) Explain the use of tapping in transformer?

ANS: Taps are provided on some transformers on the high voltage winding to correct for high or low voltage conditions, and still deliver full rated output voltages at the secondary terminals. Standard tap arrangements are at two and one-half and five percent of the rated primary voltage for both high and low voltage conditions. For example, if the transformer has a 480 volt primary and the available line voltage is running at 504 volts, the primary should be connected to the 5% tap above normal in order that the secondary voltage be maintained at the proper rating.

Qu(5) What is the Difference between "Insulating", "Isolating", and "Shielded Winding" Transformers?

ANS: Insulating and **Isolating transformers** are identical. These terms are used to describe the isolation of the primary and secondary windings, or insulation between the two. A shielded winding transformer, on the other hand, is designed with a metallic shield between the primary and secondary windings, providing a safety factor by grounding. Autotransformers is the only type whose primary and secondary are connected to each other electrically, are not of the insulating or isolating variety.

Qu(6) Can Transformers be operated at voltages other than Nameplate rating voltages?

ANS: In some cases, transformers can be operated at only voltages below the nameplate rated voltage not the higher one rating unless taps are provided for this purpose. When operating below the rated voltage the KVA capacity is reduced correspondingly

Qu(7) Can 50-Hz Transformers be Operated at 60 Hz & Vice versa ?

ANS: 50 Hz Transformers rated below 1 KVA can be utilized on 60 Hz service. However - transformers of 1 KVA & Larger rated at 60 HZ should not be used on 50 Hz service due to higher losses and resultant heat rise .

Qu(8) Is it possible that Transformers be used in Parallel?

ANS: **Single phase transformers** can be used in parallel only when their impedances and voltages are equal. If unequal voltages are used a circulating current exists in the closed network between the two transformers which will cause excess heating of the transformer. In addition, impedance values of each transformer must be within > 7.5 % of each other. For "example: Transformer A has impedance 4%, transformer B which is to be parallel to A must have impedance between the limits of 3.6 % & 4.4%.

Qu(9) Explain 60 Hz Transformers be Used at Higher Frequencies?

ANS: Transformers can be used at frequencies above 50 Hz. However, 50 Hz transformers will have less voltage regulation at 400 Hz than at 50 Hz. Where better regulation and smaller physical size are required. At 400 Hz requirement of core is not there.

Qu(10) What is Meant by Regulation in a Transformer?

ANS: Voltage regulation in transformers is the difference between the no load voltage and the full load voltage. This is usually expressed in terms of percentage. For example: A transformer delivers 100 volts at no load and the voltage drops to 90volts at full load, the regulation would be 10%.

Qu(11) What is meant by Temperature Rise in a Transformer?

ANS: Temperature rise in a transformer is the temperature of the windings and insulation above the existing ambient or surrounding temperature, and is determined by the insulation class used in the transformer coils.

Qu(12) Explain insulation "Class" in Transformer?

ANS: Insulation class was a popular way of referencing insulating materials in their ability to sustain long life while operating at different temperatures. Since it Is difficult and at times confusing to describe different insulations by letter designations, such as A, E, B, F & H; it is better to describe insulation as "insulation systems"

Qu(13) Is One Insulation System Better Than Another?

ANS: Not necessarily. For example: Small fractional KVA transformers use the class 105°C insulation system, which is 55°C rise. The class 150°C insulation system, which is 80° C rise, has generally been superseded by a class 185° C insulation system, which is 115° C rise. Medium KVA size transformers, approximately 371/2 KVA and larger, are generally manufactured using a 220° C insulation system, which is 150° C rise. All of these insulation systems from 105° C through 220° C will normally have approximately the same number of years operating life. A well designed transformer, observing these temperature limits, should have a life expectancy of approximately 20-25 years.

Qu(14) Why Should Dry Type Transformers Never be Overloaded?

ANS: Overloading of a transformer results in excessive temperature. This excessive temperature causes overheating which will result in rapid deterioration of the insulation and cause complete failure of the transformer coils.

Qu(15) Are Temperature Rise and Actual Surface Temperature Related?

ANS: No. This can be compared with an ordinary light bulb. The filament temperature of a light bulb can exceed 2000 degrees, yet the surface temperature of the bulb is low enough to permit touching with bare hands.

Qu(16) What is Meant by "Impedance" in Transformers?

ANS: Impedance is the current limiting characteristic of a transformer and is expressed in percentage.

Qu(17) Why is Impedance Important?

ANS: It is used for determining the interrupting capacity of a circuit breaker or fuse employed to protect the primary of a transformer.

Qu(18) Can Single Phase Transformers be Used for Three Phase Applications?

ANS: Yes. Three phase transformers are sometimes not readily available whereas single phase transformers can generally be found in stock. Three single phase transformers can be used in delta connected primary and wye or delta connected secondary. They should never be connected wye primary to wye secondary, since it will result in unstable Secondary voltages. The equivalent three phase capacity when properly connected of three single phase transformers is three times the nameplate rating of each single phase transformer, or example: Three 10 KVA single phase transformers will accommodate a 30 KVA three phase load

Qu(19) What is ZIG ZAG Grounding Transformer

ANS: Three Single Phase Transformers can be connected to have a three phase Zig Zag Transformer. This system can be used for either grounding or developing a fourth WIRE from a three phase neutral. An example would be to change a 480 V three phase three wire system to a 480V/277 V three phase four wire system.

Qu(20) What Color is generally used for Dry Type Transformers?

ANS: Normally light gray is used on: Single phase 50 VA through 25 KVA and on three phases 3 KVA through 15 KVA. Light gray is used with a pleasing blue-gray on the side and top covers on Single phase 37 ½ KVA through 250 KVA and three phase 20 KVA through 750 KVA.

Qu(21) How Do You Select a Transformer to Operate in an Ambient Higher Than 40° Centigrade?

ANS: If the 24 hour average ambient does not exceed 40° C, standard transformers can be used. When the ambient exceeds 40° C use the following chart for de-rating standard transformers.

Maximum Ambient	Maximum Percentage of Loading
40°	C100%
50°	C92%
60°	C85%
70° C	78%

Instead of ordering custom built transformers to operate in ambient higher than 40° C, it is more economical to use a standard transformer of a larger KVA rating.

Qu(22) Can Transformers be reconnected as Autotransformers to Increase Their KVA Rating?

ANS: Several of standard single phase transformers can be connected as autotransformers. The KVA capacity will be greatly increased when used as an Autotransformer, in comparison to the nameplate KVA as an insulating transformer. Examples of autotransformer applications are changing 600 volts to 480 volts in either single phase or three phase; changing 480 volts to 240 volts single three phase or vice versa; or the developing of a fourth wire (neutral) from a 480 volt three phase three wire system or obtaining 277 volts single phase.

This voltage is normally used for operating fluorescent lamps or similar devices requiring 277 volts.

Qu(23) What is Corona and how does it affect Dry Type Transformers?

ANS: Corona a type of localized discharge resulting from transient gaseous ionization in insulation systems when the voltage stress exceeds a critical value." In a dry type transformer, part of the insulation system is air, which is referred to in the definition as gas. If a transformer is not designed properly and the insulation is overworked, it will result in the development of ionization of the air which in turn causes deteriorating affects on the insulation of the transformer. This will result in an extremely short life of the transformer. Now day's transformers are designed so that they are corona-free even at voltages considerably higher than recommended operating voltages. The reason is, if a momentary transient voltage occurs and causes corona inception, the design of the transformer is such that the corona extinction voltage level is appreciably higher than the operating voltage level; therefore, corona will not be present during operation.

Qu(24) What is BIL and how does it apply to Transformers?

ANS: BIL is an abbreviation for Basic Impulse Level. Impulse tests are dielectric tests that consist of the application of a high frequency steep wave front voltage between windings, and between windings and ground. The Basic Impulse Level of a transformer is a method of expressing the voltage surge (lightning, switching surges, etc.) that a transformer will tolerate without breakdown. All transformers manufactured for 600 volts and below should withstand BIL rating, which is 10 KV. This assures the user that he will not experience breakdowns when his system is properly protected with lightning arrestors or similar surge protection devices.

Qu(25) What is Polarity, when associated With a Transformer?

ANS: Polarity is the instantaneous voltage obtained from the primary winding in relation to the secondary winding. Transformers 600 volts and below are normally connected in additive polarity that is, the terminals of the high voltage and low voltage windings on the left hand side are connected together, This leaves one high voltage and one low voltage terminal unconnected. When the transformer is excited, the resultant voltage appearing across a voltmeter will be the sum of the high and low voltage windings. This is useful when connecting single phase transformers in parallel for three phase operations. Polarity is a term used only with single phase transformers.

Qu(26) What is meant by Indoor or Outdoor Transformers?

ANS: Indoor transformers are ordinarily defined as transformers suitable for indoor operation only. Outdoor transformers are intended for indoor or outdoor operation. However, it is desirable to supply some protection for the transformer such that it will be shielded from direct exposure to rain, snow, or submersion in water.

Qu(27) Will a Transformer Change Three Phase to Single Phase?

ANS: A transformer will not act as a phase changing device when attempting to change three phase to single phase. There is no way that a transformer will take three phase in and deliver single phase out while at the same time presenting a balanced load to the three phase supply system. There are, however, circuits available to change three phase to two phase or vice versa using standard dual wound transformers.

Qu(28) Range of Dry Type Transformers?

ANS: Designer manufacture the most complete line available of boost-buck transformers for applications ranging from 80 to 520 volts single or three phase.

Further complete engineering and manufacturing facilities are available for custom designed transformers fractional through 5000 KVA ratings, low voltage through 15 KV. Some of the applications in which these custom built transformers are used are instrumentation, telecommunications, computer peripherals, rectifiers, reactors, oven and heating applications, and various others to match the customer's exact specifications. These transformers are available in various temperature rise and insulation systems, along with possible convection, forced air, water, or oil cooling as required by the customer.

Qu(29) How Do You Size a Transformer for Operating a DC Motor?

ANS: There are no straight forward simple formulas which can be used in sizing transformers to operate DC motors. However, the transformer size can be calculated accurately when the circuit is known for rectifying the AC to DC for operating the motor. There are a great variety of circuits now in common use for changing AC to DC. One of the more common circuits is the three phase full wave bridge circuit employing three SCR's and three diodes.

Qu(30) What is Meant by "Balanced Loading" on Single Phase Transformer Applications?

ANS: Since most single phase transformers have a secondary voltage of 120/240, they will be operated as a three wire system. Care must be taken in properly distributing the load as the transformer secondary consists of 2 separate 120 volt windings. Each 120 volt winding is rated at one-half the nameplate KVA rating. For example: A 10 KVA transformer, 120/240 volt secondary is to service an 8 KVA load at 240 volts and two 1 KVA loads at 120 volts each.

If the incorrect method is used, winding A will be loaded at 6 KVA, and winding B will be loaded at 4 KVA. These do total 10 KVA but, since each winding is only rated at 5 KVA (V2 of nameplate rating), we have an overloaded transformer and a certain failure.

Qu(31) What about Balanced Loading on Three Phases?

ANS: Each phase of a three phase transformer must be considered as a single phase transformer when determining loading. For example: A 45 KVA three phase transformer with a 208Y/120 volt secondary is to service 4 loads at 120 volts single phase each. These loads are 10 KVA, 5 KVA, 8 KVA, and 4 KVA.

Qu(32) What are the advantages and disadvantages of Induction Motors?

Ans: **Advantages:**

(i)   It is simple and rugged in construction

(ii) It is relatively cheap

(iii) Induction motors require less maintenance

(iv) Induction motor has high-efficiency and reasonably good power factor

(v)   3-phase induction machines are self-starting

Disadvantages:

(i)   Induction motors are essentially a constant speed motor and its speed cannot be changed easily

(ii) Its starting torque is inferior to dc shunt motor

Qu(33) What is the condition for maximum torque in induction motor?

Ans: Starting torque will be maximum when the rotor resistance / phase is equal to standstill rotor reactance / phase

Qu(34) Slip ring induction motor advantages and disadvantages compared to squirrel cage motors?

Ans: **Advantages:**

- High starting torque with low starting current

- Smooth acceleration under heavy loads

- No abnormal heating during starting

- Good running characteristics after external rotor resistances are cut out

- Adjustable speed

Disadvantages:

- The initial and maintenance costs are greater than those of squirrel cage motors

- The speed regulation is poor when run with resistance in the rotor circuit

Qu(35) Methods to control speed of Wound Rotor Motors?

Ans: The speed of wound rotor motors is changed by changing the slip of the motor. This can be achieved by:

- Varying the stator line voltage

- Varying the resistance in the rotor circuit

- Inserting and varying a foreign voltage source in the rotor circuit

Qu(36) Explain how Torque-Slip Characteristics vary when adding resistance to rotor circuit?

Ans: The addition of resistance to the rotor circuit does not change the value of maximum torque but it only changes the value of the slip at which the maximum torque occurs.

Qu(37) Disadvantages of Star-Delta Starting of Induction motor?

Ans: In Star-Delta starting induction motor stator is connected in star connection for starting after picking up speed it is connected to delta connection. When induction motor is connected in star connection stator phase voltage reduced by $1/(3^{1/2})$ times the line voltage. This also results in reduced starting torque (1/3 times compared to delta connection).

Qu(38) .Name the two important parameters that attribute to efficiency of electricity use by AC induction Motors?

Ans: The important parameters that attribute to efficiency of electricity use by AC induction Motors are

1. Efficiency of the motor

2. Power Factor

Qu(39) Define percentage unbalance in voltage?

Ans: Percentage unbalance in voltage is defined as $[(V_{max} - V_{avg})/V_{avg}] \times 100$, where V_{max} and V_{avg} is the largest and the average of the three phase voltages respectively.

Qu(40) Give the reason why motors operate in star mode for under loaded motors?

Ans:-For motors which consistently operate at loads below 50 % of rated capacity, an inexpensive and effective measure might be to operate in star mode. A change from the standard delta operation to star operation involves re-configuring the wiring of the three phases of power input at the terminal box. Operating in the star mode leads to a voltage reduction by a factor of '$\sqrt{3}$'. Motor output falls to one-third of the value in the delta mode, but performance characteristics as a function of load remain unchanged. Thus, full-load operation in star mode gives higher efficiency and power factor than partial load operation in the delta mode. However, motor operation in the star mode is possible only for applications where the torque-to-speed requirement is lower at reduced load.

Qu(41) Write down applications of constant torque and variable torque loads.

Ans: Constant Torque Loads: Conveyors, rotary kilns, constant displacement pumps Variable torque loads: Centrifugal pumps and fans.

Qu(42) Write down the important parameters that influence the motor selection?

Ans: The following parameters influence the motor selection:

(i) Torque requirement/load characteristics

(ii) Ambient operating conditions

(iii) Anticipated switching frequency

(iv) Reliability

(v) Inventory

(vi) Price

(vii) Efficiency

Qu(43)What down different types of losses in any motor?

Ans: The different losses in any motor are shown below:

Fixed Loses – core loss, friction & windage loss.

Variable Losses - Copper & Stray Losses

Qu(44) Why 'induction motors' are so popular over all types of motors?

Ans: (i) Low cost (compared with DC) and Wide availability

(ii) Low maintenance - no brushes or commutator

(iii)Rugged design - can be used in harsh environments

(iv)Low inertia rotor designs

(v)High electrical efficiency

(vi)Wide speed ranges

(vii)No separately-powered field windings

Qu(45) How size of the capacitor rating decided for an induction motor?

Ans:The size of capacitor required for a particular motor depends upon the no-load reactive kVA (kVAR) drawn by the motor, which can be determined only from no-load testing of the motor. In general, for full loading operating motor, the capacitor selected to not exceed 90 % of the no-load kVAR of the motor. (Higher capacities could result in over-voltages and motor burn-outs).

Qu(46) Name the methods for correcting poor power factor in motors?

Ans: The following are strategies for correcting power factor in motors:

1. Minimize operation of idling or lightly loaded motors

2. Ensuring correct supply of rated voltage and phase balance

3. Installing capacitors to decrease reactive power loads

Qu(47) Explain different types of losses in the Induction Motor?

Ans:- 1. No-load Losses: These losses are independent of load .it remain constant throughout the operation of machine.

2. Load dependent Losses: Vary as function of motor loading

The losses in a motor are of two types such as fixed i.e. independent of load on the motor and the other variable i.e. dependent on the load.

Fixed losses consist of Iron loss and mechanical loss (friction and windage loss). The iron loss vary with the material and geometry and with input voltage whereas friction and windage losses are caused by friction of the motor and aerodynamic losses associated with the ventilation fan and other rotating parts.

Variable losses consist of resistance losses in the stator and in the rotor. Stray losses arise from a variety of sources and are difficult to measure directly or to calculate and are generally considered proportional to the square of the rotor current.

Qu(48) What are the factors to be considered while selecting a motor?

Ans:-A. *Torque Requirement*

The primary consideration defining the motor choice for any particular application is the torque required by the load. The relationship between the maximum torque generated by the motor (break-down torque) and the torque requirements for start-up (locked rotor torque) and during acceleration periods is very important

***B*.** Sizing to Variable Load Industrial motors frequently operate under varying load conditions due to process requirements.

The optimum rating for the motor is selected on the basis of the load duration curve for the particular application. Thus, rather than selecting a motor of high rating that would operate at full capacity for only a short period, a motor would be selected with a rating slightly lower than the peak anticipated load and would operate at overload for a short period of time. Since operating within the thermal capacity of the motor insulation is of greatest concern in a motor operating at higher than its rated load, the motor rating is selected as that which would result in the same temperature rise under continuous full-load operation as the weighted average temperature rise over the actual operating cycle.

Qu(49) Write down the effects of harmonics on motor operation and performance?

Ans:-Harmonics increase motor losses, and can adversely affect the operation of sensitive auxiliary equipment. The non-sinusoidal supply results in harmonic currents in the stator which increases the total current drawn. In addition, the rotor resistance (or more precisely, impedance) increases significantly at harmonic frequencies, leading to less efficient operation. Also, stray load losses can increase significantly at harmonic frequencies. Overall motor losses increase by about 20% with a six-step voltage waveform compared to operation with a sinusoidal supply. In some cases the motor may have to be de-rated as a result of the losses. Alternatively, additional circuitry and switching devices can be employed to minimize losses. Instability can also occur due to the interaction between the motor and the converter. This is especially true of motors of low rating, which have low inertia. Harmonics can also contribute to low power factor.

Qu(50) Define winding factor?

Ans;-The winding factor K_d is defined as the ratio of phasor addition of emf induced in all the coils belonging to each phase winding to their arithmetic addition

Qu(51) Why are Alternators rated in KVA and not in KW ?

Ans:-The continuous power rating of any machine is generally defined as the power the machine or apparatus can deliver for a continuous period so that the losses incurred in the machine gives rise to a steady temperature rise not exceeding the limit prescribed by the insulation class. copper loss, occurring in the 3 -phase winding which depends on I^2 R. As the current is directly related to apparent - power delivered by the generator, the Alternators have only their apparent power in VA/KVA/MVA as their power rating.

Qu(52) What are the causes of changes in voltage in Alternators when loaded?

Ans:- (i)Voltage variation due to the resistance of the winding, R

(ii) Voltage variation due to the leakage reactance of the winding.

(iii) Voltage variation due to the armature reaction effect, X_a.

Qu(53) What is meant by load angle of an Alternator?

Ans: The phase angle introduced between the induced emf phasor E and terminal voltage phasor V, during the load condition of an Alternator is called load angle .

Qu(54) An Alternator is found to have its terminal voltage on load condition more than that on no load. What is the nature of the load connected?

Ans:- The nature of the load is of leading power factor , load consisting of resistance and capacitive reactance.

Qu(55) Define the term voltage regulation of Alternator.
Ans:- The voltage regulation of an Alternator is defined as the change in terminal voltage from no-load to load condition expressed as a fraction or percentage of terminal voltage at load condition ; the speed and excitation conditions remaining same.

Qu(56) What is the necessity for predetermination of voltage regulation?
Ans:- For Alternators of large power and voltage ratings conducting load test is not possible. Hence other indirect methods of testing are used and the performance like voltage regulation then can be predetermined at any desired load currents and power factors.

Qu(57) Name the various methods for predetermining the voltage regulation of 3-phase Alternator?
Ans:(1)-Synchronous impedance /EMF method
(2)Ampere-turn / MMF method .
(3)Poitier / ZPF method

Qu(58) How does the ampere-turn method differ from synchronous impedance method?
Ans:- The ampere-turn /MMF method is the converse of the EMF method in the sense that instead of having the phasor addition of various voltage drops/EMFs, here the phasor addition of MMF required

for the voltage drops are carried out. Further the effect of saturation is also taken care of.

Qu(59) What are the test data required for voltage regulation of an Alternator by MMF method?

Ans (i) Effective resistance per phase of the 3-phase winding R.

(ii) Open circuit characteristic (OCC) at rated speed/frequency.

(iii) Short circuit characteristic (SCC) at rated speed/frequency.

Qu(60) Why is the MMF method of estimating the voltage regulation considered as the optimistic method?

Ans: Compared to the EMF method, MMF method, involves more number of complex calculation steps. Further the OCC is referred twice and SCC is referred once while predetermining the voltage regulation for each load condition. Reference of OCC takes care of saturation effect. As this method requires more effort, the final result is very close to the actual value. Hence this method is called optimistic method.

Qu(61) What are thr condition for connecting two alternators in parallel?

Ans: (i)The terminal voltage magnitude of the incoming Alternator must be made equal to the existing Alternator or the bus-bar voltage magnitude

(ii) The phase sequence of the incoming Alternator voltage must be similar to the bus-bar voltage.

(iii) The frequency of the incoming Alternator voltage must be the same as the bus-bar voltage.

Qu(62) How do the synchronizing lamps indicate the correctness of phase sequence between existing and incoming Alternators?

Ans:- The correctness of the phase sequence can be checked by looking at the three sets of lamps connected across the 3-pole of the synchronizing switch. If the lamps grow bright and dark in unison it is an indication of the correctness of the phase sequence. If on the other hand, they become bright and dark one after the other, connections to any two machine terminals have to be interchanged after shutting down the machine

Qu(63) Why synchronous generators are to be constructed with more synchronous reactance and negligible resistance?

Ans: The presence of more resistance in the Synchronous generators will resist or oppose their synchronous operation. More reactance in the generators can cause good reaction between the two and help the generators to remain in synchronism in spite of any disturbance occurring in any one of the generators

Qu(64)What are the various factors that affect the load sharing in parallel operating generators?

Ans: (i) Prime-mover characteristic/input

(ii) Excitation level

(iii) Percentage synchronous impedance and its R/X ratio

Qu(65) How does the change in prime mover input affect the load sharing?

Ans: An increase in prime-mover input to a particular generator causes the active power shared by it to increase and a corresponding decrease in active-power shared by other generators. The change in reactive power sharing is less appreciable. The frequency of the bus-bar voltage will also subjected to slight increase in value.

Qu(66) How does change in excitation affects the load sharing?

Ans: The decrease in excitation in one generator causes the reactive power shared by it to decrease and a corresponding increase in reactive-power shared by other generators. The change in active-power sharing is less appreciable. There will be a slight decrease in terminal voltage magnitude also

Qu(67) What steps are to be taken before disconnecting one Alternator from parallel operation?

Ans: (i) The prime-mover input of the outgoing generator has to be decreased and that of other generators has to be increased and by this the entire active-power delivered by the outgoing generator is transferred to other generators

Qu(68) Why large size Synchronous machines are constructed with rotating field system type ?

Ans: Advantages of the rotating field system type construction of Synchronous machines

(i) The relatively small amount of power, about 2%, required for field system via slip-rings and brushes.

(ii) For the same air gap dimensions, which is normally decided by the KVA rating, more space is available in the stator part of the machine for providing more insulation to the system of conductors.

(iii) Insulation to stationary system of conductors is not subjected to mechanical stresses.

(iv) It is easy to provide cooling arrangement for a stationary system of conductors.

Qu(69) Classify alternators?

Ans:- **According to type of field system**

(i) Stationary field system type
(ii) Rotating field system type

According to shape of field system

(i) Salient pole type
(ii) Smooth cylindrical type

Qu(70) Why cylindrical Alternators are used in steam turbines?

Ans : Steam turbines are found to operate at fairly good efficiency only at high speeds. The high speed operation of rotors tends to increase mechanical losses and so the rotors should have a smooth external surface. Hence, smooth cylindrical type rotors with less diameter and large axial length.

Qu(71) Why salient pole type Synchronous generators are used in Hydro-electric plants ?

Ans:As the speed of operation is low for hydro turbines used in Hydro-electricplants, salient pole type Synchronous generators are used. These allow better ventilation and also have other advantages over smooth cylindrical type rotor. Salient pole type rotors with large diameter and small axial length.

Qu(72) A 4-pole 415 V 3-phase, 50 Hz induction motor runs at 1440 RPM at .88 pf lagging and delivers 10.817 kW. The stator loss is 1060 W, and friction & windage losses are 375 W. Find

(i). Slip

(ii). Rotor Copper loss

(iii). Line current

(iv). Efficiency.

solution: Supply frequency (f) =50Hz

No. of poles (P) =4 Synchronous speed (Ns) =120f/P=1500 RPM

Actual speed (Nm)=1440 RPM

A. Slip(s) = (Ns – Nm)/ Ns = **0.04pu**

B. Motor output =10817W

Stator Cu loss =1060W

Friction & windage =375W

Motor Output= Rotor Input – Rotor Copper loss – Friction & windage loss (1)

We know

Rotor Input = Rotor Copper loss/ slip

Substituting in (1)

Rotor Copper Loss = (Motor output + Friction & windage loss) x slip/ (1- slip)

Therefore, Rotor Copper loss= (10817+375) x 0.04/ (1-0.04)

Rotor Copper loss= **466.33W**

C. Motor input = Rotor input + Stator loss

Rotor input = Rotor Copper loss/slip = 466.33/0.04 = 11658W Motor input = 11658 + 1060 = 12718W V= 415 Volts; Cos φ = 0.88lag;

Line Current =Motor input/ (1.732 x V x Cos φ) = **20.11A**

D. Efficiency = (Motor Output/Motor input) x 100%

= (10817/12718) x 100

= **85%**

Multiple Choice Questions

1 Harmonics in transformer result in

(A) Increased core losses

(B) Increased I2R losses

(C) Magnetic interference with communication circuits

(D) All of the above

Ans: D

2.The full load copper loss of a transformer is 1600W. At half-load the copper loss will be

(A) 6400W

(B) 1600W

(C) 800W

(D) 400W

Ans:D

3. Power transformers are generally designed to have maximum efficiency around

(A) No load

(B) Half load

(C) Near full load

(D) 10% overload

Ans: C

4.Two transformers are connected in parallel. These transformers do not have equal percentage impedance which results

(A) Short-circuiting of the secondary's

(B) Power factor of one of the transformers is leading while that of the other lagging

(C) Transformers having higher copper losses will have negligible core losses

(D) Loading of the transformers not in proportion to their kVA ratings.

Ans:D

5.The changes in volume of transformer cooling oil due to variation of atmospeheric temperature during day and night is taken care of by which part of transformer?

(A) Conservator

(B) Breather

(C) Bushings

(D) Buchholz relay

Ans:A

6.The transformer laminations are insulated from each other by

(A) Mica strip

(B) Thin coat of varnish

(C) Paper

(D) Any of the above

Ans:B

7.Which type of winding is used in 3 phase shell type transformer?

(A) Circular type

(B) Sandwich type

(C) Cylindrical type

(D) Rectangular type

Ans:B

8.During open circuit **test** of a transformer

(A) Primary is supplied rated voltage

(B) Primary is supplied full load current

(C) Primary is supplied current at reduced voltage

(D) Primary is supplied rated KVA

Ans:A

9. Which of the following is not standard voltage for power supply in India

(A) 11kV

(B) 33kV

(C) 66 kV

(D) 122 kV

Ans:D

10. The low voltage winding of a 400/230 volt, 1-phase, 50Hz transformer is to be connected to a 25Hz supply in order to keep the magnetization current at the same level as that for normal 50Hz supply at 25Hz the voltage should be [Gate 1997]

(A) 230V

(B) 460V

(C) 115V

(D) 65V

Ans:C

11.The synchronous speed 6 pole motor operating at50Hz frequency_____.

(A) 1500

(B) 1000

(C) 750

(D) 3000

Ans :B

12. The primary current in the current transformer is dictated by [Gate 1997]

(A) The secondary burden

(B) The core of the transformer

(C) The load current

(D) None of the above

Ans:A

13.A Buchholz relay can be installed on

(A) Auto-transformers

(B) Air-cooled transformers

(C) Welding transformers

(D) Oil cooled transformers

Ans:D

14.The chemical used in breather is

(A) Asbestos fiber

(B) Silica sand

(C) Sodium chloride

(D) Silica gel

Ans: D

15. Material used for construction of transformer core is usually

(A) 0.4mm to 0.5 mm

(B) 4mm to 5 mm

(C) 14mm to 15 mm

(D) 25mm to 40 mm

Ans:D

16. Helical coils can be used on

(A) Low voltage side of high kVA transformers

(B) High frequency transformers

(C) High voltage side of small capacity transformers

(D) High voltage side of high kVA rating transformers

Ans: A

17. The transformer ratings are usually expressed in terms of

(A) Volts

(B) Amperes

(C) kW

(D) KVA

Ans: D

18. The value of flux involved in the emf equation of a transformer is

(A) RMS value

(B) RMS value

(C) Maximum value

(D) Instantaneous value

Ans: C

19.The main advantage of auto transformer over a two winding transformer is

(A) Hysteresis losses are reduced

(B) Saving in winding material

(C) Copper losses are negligible

(D) Eddy losses are totally eliminated

Ans:B

20. During short circuit **test** iron losses are negligible because

(A) The current on secondary side is negligible

(B) The voltage on secondary side does not vary

(C) The voltage applied on primary side is low

(D) Full-**load** current is not supplied to the transformer

Ans: C

21. Which of the following properties is not necessarily desirable in the material for transformer core?

(A) Mechanical strength

(B) Low hysteresis loss

(C) High thermal conductivity

(D) High permeability

Ans: C

22. The main reason for generation of harmonics in a transformer could be

(A) Fluctuating load

(B) Poor insulation

(C) Mechanical vibrations

(D) Saturation of core

Ans: D

23. In a transformer the energy is conveyed from primary to secondary

(A) Through cooling coil

(B) Through air

(C) By the flux

(D) None of the above

Ans:C

24. Which loss is not common between a transformer and rotating machines?

(A) Eddy current loss

(B) Copper loss

(C) Windage loss

(D) Hysteresis loss

Ans:A

25. No load test on a transformer is carried out to find

(A) Copper loss

(B) Magnetizing current

(C) Magnetizing current and loss

(D) Efficiency of the transformer

Ans:C

26. Sumpner's test is conducted on transformers to find

(A) Temperature

(B) Stray losses

(C) All-day efficiency

(D) None of the above

Ans:A

27. The efficiency of a transformer will be maximum when

(A) Copper losses = hysteresis losses

(B) Hysteresis losses = eddy current losses

(C) Eddy current losses = copper losses

(D) Copper losses = iron losses

Ans: D

28.The purpose of providing an iron core in a transformer is to?

(A) Provide support to windings

(B) Reduce hysteresis loss

(C) Decrease the reluctance of the magnetic path

(D) Reduce eddy current losses

Ans: C

29. The highest voltage for transmitting electrical power in India is?

(A) 33kV

(B) 66kV

(C) 132kV

(D) 400kV

Ans:D

30. The function of conservator in a transformer is

(A) To protect against internal fault

(B) To reduce copper as well as core losses

(C) To cool the transformer oil

(D) To take care of the expansion and contraction of transformer oil due to variation of temperature of surroundings

Ans: D

31. The core used in high frequency transformer is usually

(A) Copper core

(B) Cost iron core

(C) Air core

(D) Mid steel core

Ans: C

32. Cross over windings are used in

(A) Low voltage side of high kVA rating transformers

(B) Current transformers

(C) High voltage side of high kVA rating transformers

(D) High voltage side of low kVA rating transformers

Ans: D

33.A transformer oil must be free from

(A) Sludge

(B) Odor

(C) Gases

(D) Moisture

Ans:D

34. Silicon steel used in laminations mainly reduces

(A) Hysteresis loss

(B) Eddy current losses

(C) Copper loss

(D) All of the above

Ans: A

35.In case there are burrs on the edges of the laminations of the transformer, it is likely to result in

(A) Vibrations

(B) Noise

(C) Higher eddy currents loss

(D) Higher hysteresis loss

Ans: C

36. Which of the following is not a part of transformer installation?

(A) Conservator

(B) Breather

(C) Buchholz relay

(D) Exciter

Ans:D

37. In a transformer the toppings are generally provided on

(A) Primary side

(B) Secondary side

(C) Low voltage side

(D) High voltage side

Ans: C

38. The effect of increasing the length of the air gap in an induction motor will increase

(A) Power factor

(B) Speed

(C) Magnetizing current

(D) Air-gap flux

Ans: C

39.the difference between the synchronous speed and the actual speed of an induction motor is known as

(A) Regulation

(B) Back lash

(C) Slip

(D) Lag

Ans:C

40.Rotating magnetic field is produced in a....

(A) Single - phase induction motor

(B) Three phase induction motor

(C) DC series motor

(D) AC series motor

Ans: B

41.The stator core of the induction motor is made of

(A) Laminated cast iron

(B) Mild steel

(C) Silicon steel stampings

(D) Soft wood

Ans:C

42.Star- delta starter of an induction motor

(A) Inserts resistance in rotor circuit

(B) Inserts resistance in stator circuit

(C) Applies reduced voltage to rotor

(D) Applies reduced voltage to stator

Ans: D

43. The thrust developed by a linear induction motor depends on

(A) Synchronous speed

(B) Rotor input

(C) Number of poles

(D) Both A and B

Ans: D

44.If an induction machine is run at above synchronous speed, it acts as [Gate 1997]

(A) A synchronous motor

(B) An induction generator

(C) An induction motor

(D) None of the above

Ans: B

45. In a synchronous motor, damper winding is provided to

(A) Stabilize rotor motion

(B) Suppress rotor oscillations

(C) Develop necessary starting torque

(D) Both B and C

Ans: D

46. Synchronous capacitor is

(A) An ordinary static capacitor bank

(B) An over excited synchronous motor driving mechanical load

(C) An over excited synchronous motor running without mechanical load

(D) None of the above

Ans:C

47. A synchronous machine is called as doubly excited machine because

(A) It can be over excited

(B) It has two sets of rotor poles

(C) Both its rotor and stator are excited

(D) It needs twice the normal exciting current

Ans: C

48. If the field of a synchronous motor is under excited, the power factor will be

(A) Lagging

(B) Leading

(C) Unity

(D) More than unity

Ans: A

49.The direction of rotation of a synchronous motor can be reversed by reversing

(A) Current to the field winding

(B) Supply phase sequence

(C) Polarity of rotor poles

(D) None of the above

Ans: B

50.A synchronous motor connected to infinite bus bars has at constant full-load, 100% excitation and unity pf. On changing the excitation only, the armature current will have

(A) Leading pf with under-excitation

(B) Leading pf with over excitation

(C) Lagging pf with over excitation

(D) No change of pf

Ans: B

51.The angle between the synchronous rotating stator flux and rotor poles of a synchronous motor is

(A) Synchronizing angle

(B) Torque angle

(C) Power factor angle

(D) Slip angle

Ans:B

52.In a synchronous machine when the rotor speed becomes more than the synchronous speed during hunting, the damping bars develop

(A) Synchronous motor torque

(B) DC motor torque

(C) Induction motor torque

(D) Induction generator torque

Ans: D

53. When load on a synchronous motor is increased its armature current is increased provided it is

(A) Normally excited

(B) Over excited

(C) Under exciter

(D) All of the above

Ans:D

54.In order to reduce the harmonics in the emf generated in an alternator

(A) Slots are skewed

(B) Salient pole tips are chamfered

(C) Winding is well distributed

(D) all of the above.

Ans: D

55. The maximum power in a synchronous machine is obtained when the load angle is

(A) 0°

(B) 85°

(C) 120°

(D) 135°.

Ans: B

56.The emf generated due to nth harmonic component of flux in an alternator will be

(A) n times the fundamental emf

(B) same as fundamental emf

(C) less than the value of fundamental emf.

Ans: C

57. Synchronizing torque comes into operation under all of the following cases EXCEPT

(A) Phase difference between two voltages

(B) Frequency difference between two voltages

(C) Voltage difference between two voltages

(D) Reduction in exciting current in one of the alternators.

 Ans: D

58. Unbalanced 3-phase stator currents cause

(A) Double frequency currents in the rotor

(B) Healing of rotor

(C) Vibrations

(D) all of the above.

Ans: D

59. In large generators protection provided against external faults is

(A) Biased differential protection

(B) Sensitive earth fault protection

(C) Inter-turn fault protection

(D) all of the above.

Ans: D

60. Pitch factor is the ratio of the emfs of

(A) Short pitch coil to full pitch coil

(B) Full pitch winding to concentrated winding

(C) Full pitch winding to short pitch winding

(D) Distributed winding to full pitch winding.

Ans: A

61. In an alternator if the winding is short pitched by 50 electrical degrees, its pitch factor will be

(A) 1.0

(B) 0.866

(C) 0.75

(D) 0.50.

Ans:B

62. If a single phase alternator has 8 slots per pole uniformly speed, but the winding is arranged with the middle two left empty, the breadth coefficient will be

(A) 0.99

(B) 0.88

(C) 0.67

(D) 0.53.

Ans: D

63. Two alternators are running in parallel. If the field of one of the alternator is adjusted, it will

(A) Reduce its speed

(B) Change its load

(C) Change its power factor

(D) Change its frequency.

Ans: C

64 A generator is operating by itself supplying the system loads. The reactive power supplied by the generator will

(A) Depend on prime mover rpm

(B) Depend on type of insulation used

(C) Depend on the amount demanded by the load

(D) Depend on inter-coil inductance.

Ans: C

65. Which of the following part plays important role in over speed protection of a generator ?

(A) Over current relay

(B) Alarm

(C) Differential protection

(D) Governor.

Ans: D

66. Which relays comes into operation in the event of the failure of prime mover connected to the generator?

(A) Reverse power relay

(B) Differential relay

(C) Buchholz relay

(D) None of the above.

Ans: A

67The windings for an alternator are

I. 36 slots, four poles, span 1 to 8

II. 72 slots, six poles, span 1 to 10

III. 96 slots, six poles, span 1 to 12.

The windings having pitch factors of more than 0.9 are

(A) I and II only

(B) II and III only

(C) I and II only

(D) I, II and III.

Ans: A

68 A 500 kVA, 2300 volt three phase star connected alternator has a full load armature-resistance drop per phase of 50 volts and a combined armature reactance plus armature-reaction drop of 500 volts per phase.

The percent regulation of the alternator at unity power factor is

(A) 1.05

(B) 10.5

(C) 21.5

(D) 27.5.

Ans:B

69. In above question The percent regulation of the alternator at 0.8 power factor leading is

(A) 13.2

(B) 26.4

(C) - 26.4

(D) - 13.2.

Ans: D

70. The imaginary or fictitious part of synchronous reactance takes care of

(A) Armature reaction

(B) Voltage regulation

(C) Inductive reactance

(D) All of the above.

Ans: A

71. In an alternator, the use of short pitch coils of 160° will indicate the absence of

(A) Third harmonic

(B) Fifth harmonic

(C) Seventh harmonic

(D) Ninth harmonic.

Ans: D

72. When a generator designed for operation at 60 Hz is operated at 50 Hz

(A) Operating voltage decreased to (50/60) of its original value

(B) Operating voltage decreased to (50/60)2 of its original value

(C) KVA rating increased to (60/50) of the rated value

(D) The generator will not take any load.

Ans: A

73 The voltage of field system for an alternator is usually

(A) Less than 200 V

(B) Between 200 V and 440 V

(C) More than 400 V

(D) More than 1 kV.

Ans: A

74 Maximum electric power output of a synchronous generator is

(A) Xs / VtEf

(B) V2t / Xs

(C) E2f / Xs

(D) VtEf / Xs

Ans: D

75 The electrical angle between the field axis and axis of armature reaction of a loaded synchronous generator with armature current lagging behind the excitation emf by ψ is

(A) ψ - 90

(B) ψ + 90

(C) 90 - ψ

(D) ψ + 180.

Ans: B

76Two synchronous generators G1and G2 are equally sharing the KVAR of the load while operating in parallel. Keeping the terminal voltage fixed in order to shift part of the KVAR load from G2 to G1

(A) The field current of G1 is lowered

(B) The field current of G2 is raised

(C) The field current of G1 is raised and of G2 lowered

(D) The field current of G1 is lowered and of G2 is raised.

Ans: C

77 A synchronous generator is operating with excitation adjusted for unity power factor current at constant load. When on increasing the excitation the power factor

(A) Will lag

(B) Will lead

(C) Will become zero

(D) All of the above.

Ans: A

78 On changing the speed of an alternator from 4000 rpm to 2000 rpm, the generated emf phase will become

(A) 1/4

(B) 1/2

(C)1/3

(D)1/5

Ans:B

79Synchronous motor can operate at

(A) Lagging power factor only

(B) Leading power factor only

(C) Unity power factor only

(D) Lagging, leading and unity power factor only.

Ans:D

80 An unexcited single phase synchronous motor is

(A) Reluctance motor

(B) Repulsion motor

(C) Universal motor

(D) AC series motor.

Ans:A

81 The maximum power developed in the synchronous motor will depend on

(A) Rotor excitation only

(B) Maximum value of coupling angle

(C) Supply voltage only

(D) Rotor excitation supply voltage and maximum value of coupling angle.

Ans:D

82 In case the field of a synchronous motor is under excited, the power factor will be

(A) Leading

(B) Lagging

(C) Zero

(D) Unity.

Ans:B

83 A synchronous motor is switched on to supply with its field windings shorted on them. It will

(A) Not start

(B) Start and continue to run as an induction motor

(C) Start as an induction motor and then run as synchronous motor

(D) Start immediately.

Ans:C

84 The damping winding in a synchronous motor is generally used

(A) To provide starting torque only

(B) To reduce noise level

(C) To reduce eddy currents

(D) To prevent hunting and provide the starting torque.

Ans:D

85 The back emf set up in the stator of a synchronous motor will depend on

(A) Rotor speed only

(B) Rotor excitation only

(C) Rotor excitation and rotor speed

(D) Coupling angle, rotor speed and excitation.

Ans:B

86 A synchronous motor is a useful industrial machine on account of which of the

following reasons ?

I. It improves the power factor of the complete installation

II. Its speed is constant at all loads, provided mains frequency remains constant

III. It can always be adjusted to operate at unity power factor for optimum efficiency and economy.

(A) I only

(B) II only

(C) III only

(D) I, II and III.

Ans:D

87 Which of the following is an unexcited single phase synchronous motor ?

(A) A.C. series motor

(B) Universal motor

(C) Reluctance motor

(D) Repulsion motor.

Ans:C

88. An over excited synchronous motor draws current at

(A) Lagging power factor

(B) Leading power factor

(C) Unity power factor

(D) Depends on the nature of load.

Ans:B

89. With the increase in the excitation current of synchronous motor the power factor of the motor will

(A) Improve

(B) Decrease

(C) Remain constant

(D) Depend on other factors.

Ans:A

90. The armature current of a synchronous motor has large values for

(A) Low excitation only

(B) High excitation only

(C) Both low and high excitation

(D) Depends on other factors.

Ans:C

91. A synchronous motor is switched on to supply with its field windings shorted on themselves. It will

(A) Not start

(B) Start and continue to run as an induction motor

(C) Start as induction motor and then run as a synchronous motor.

Ans:C

92. If the field of a synchronous motor is under excited, the power factor will be

(A) Lagging

(B) Leading

(C) Unity.

Ans:A

93. When the excitation of an unloaded salient-pole synchronous motor suddenly gets disconnected

(A) The motor stops

(B) It runs as a reluctance motor at the same speed

(C) It runs as a reluctance motor at a lower speed.

Ans:A

94. The armature current of the synchronous motor has large values for

(A) Low excitation only

(B) High excitation only

(C) Both high and low excitation.

Ans:C

95. What is the ratio of no load speed to full load speed of a 200 kVA, 12 pole, 2200 V, 3 phase, 60 Hz synchronous motor?

(A) 1

(B) 1.1

(C) 1.21

(D) infinite.

Ans:A

96. If a synchronous motor drops too far behind, the power it takes from the supply also increases too much, and the armature tries to get accelerated, until it is in correct position. Sometimes, some motor overshoots the marks and then the process of acceleration-retardation continues. This phenomenon is known as

(A) Synchronization

(B) Hunting

(C) Pulling out

(D) Swinging.

Ans:B

97.The maximum value of torque that a synchronous motor, can develop without losing its synchronism, is known as

(A) Breaking torque

(B) Synchronizing torque

(C) Pull out torque

(D) Slip torque.

Ans:B

98. In a synchronous motor if the back emf generated in the armature at no load is approximately equal to the applied voltage, then

(A) The torque generated is maximum

(B) The excitation is said to be zero percent

(C) The excitation is said to be 100%

(D) The motor is said to be fully loaded.

Ans:C

99.A 3 phase, 400 V, 50 Hz salient pole synchronous motor is fed from an infinite bus and is running at no load. Now if the field current of the motor is reduced to zero

(A) The motor will stop

(B) The motor will run

(C) The motor will run at synchronous speed

(D) The motor will run at less than synchronous speed.

Ans:C

100.The purpose of embedding the damper winding in the pole face is to

(A) Eliminate hunting and provide adequate starting torque

(B) Reduce windage losses

(C) Eliminate losses on account of air friction

(D) Reduce bearing friction.

Ans:A

101. A synchronous motor is switched on to supply with its field windings shorted on themselves. It will

(A) Not start

(B) Start but continue to run as an induction motor

(C) Start as an induction motor and then run as a synchronous motor.

(D) None of the above.

Ans: B

102 The armature of a dc machine is laminated to reduce:

(A) Eddy current loss (B) Hysteresis loss

(C) Copper losses (D) friction and windage losses

Ans: A

Thinner the laminations, greater is the resistance offered to the induced e.m.f., smaller the current and hence lesser the I2R loss in the core.

103. In case of a synchronous motor we have

I. Load

II. Speed

III. DC excitation.

The magnitude of stator back emf depends on

(A) I only

(B) I and II only

(C) III only

(D) I, II and III.

Ans:C

104Out of the following methods of heating the one which is independent of supply

frequency is

(A) Electric arc heating

(B) Induction heating

(C) Electric resistance heating

(D) Dielectric heating

Ans: C

105In a stepper motor the angular displacement

(A) Can be precisely controlled.

(B) It cannot be readily interfaced with micro computer based controller.

(C) The angular displacement cannot be precisely controlled.

(D) It cannot be used for positioning of work tables and tools in NC machines.

Ans: A

106The generation voltage is usually

(A) Between 11 KV and 33 KV.

(B) Between 132 KV and 400 KV.

(C) Between 400 KV and 700 KV.

(D) None of the above.

Ans: A

107 when a synchronous motor is running at synchronous speed, the damper winding produces

(A) Damping torque.

(B) Eddy current torque.

(C) Torque aiding the developed torque.

(D) No torque.

Ans: D

108. A hysteresis motor

(A) Is not a self-starting motor.

(B) Is a constant speed motor.

(C) Needs dc excitation.

(D) Cannot be run in reverse speed.

Ans: B

109 Which of the following motors is non-self starling ?

(A) squirrel cage induction motor

(B) wound rotor induction motor

(C) synchronous motor

(D) DC series motor.

Ans :C

110. The back emf in the stator of a synchronous motor depends on

(A) speed of rotor

(B) rotor excitation

(C) number of poles

(D) flux density.

Ans:B

111.Which motor can conveniently operate on lagging as well as leading power factor ?

(A) squirrel cage induction motor

(B) wound rotor induction motor

(C) synchronous motor

(D) any of the above.

Ans:C

112. A synchronous motor working on leading power factor and not driving any mechanical, is known

(A) synchronous induction motor

(B) spinning motor

(C) synchronous condenser

(D) none of the above.

Ans:C

113.The constant speed of a synchronous motor can be changed to new fixed value by

(A) changing the applied voltage

(B) interchanging any two phases

(C) changing the load

(D) changing the frequency of supply.

Ans:D

114.The relative speed between the magnetic fields of stator and rotor under steady state operation is zero for a

(A) DC machine.

(B) 3 phase induction machine.

(C) Synchronous machine.

(D) Single phase induction machine.

Ans: All are correct

115. The current from the stator of an alternator is taken out to the external load circuit through

(A) Slip rings.

(B) Commutator segments.

(C) Solid connections.

(D) Carbon brushes.

Ans: C

116.A motor which can conveniently be operated at lagging as well as leading power factors is the

(A) Squirrel cage induction motor.

(B) Wound rotor induction motor.

(C) Synchronous motor.

(D) DC shunt motor.

Ans: C

117. The most suitable servomotor for low power applications is

(A) A dc series motor.

(B) A dc shunt motor.

(C) An ac two-phase induction motor.

(D) An ac series motor.

Ans: B

118.The size of a **conductor** used in power cables depends on the

(A) Operating voltage.

(B) Power factor.

(C) Current to be carried.

(D) Type of insulation used.

Ans: C

119.The two windings of a transformer is

(A) Conductively linked.(B) inductively linked.

(C) Not linked at all.(D) electrically linked.

Ans : B

120 A salient pole synchronous motor is running at noload. Its field current is switched off.

The motor will

(A) Come to stop.

(B) Continue to run at synchronous speed.

(C) Continue to run at a speed slightly more than the synchronous speed.

(D) Continue to run at a speed slightly less than the synchronous speed.

Ans: B

121 The d.c. series motor should always be started with load because

(A) At no load, it will rotate at dangerously high speed.

(B) It will fail to start.

(C) It will not develop high starting torque.

(D) All are true.

Ans: A

122The frequency of the rotor current in a 3 phase 50 Hz, 4 pole induction motor at full

load speed is about

(A)200 Hz.   (B) 20 Hz.

(C)2 Hz.   (D) 0.2Hz.

Ans: C

123.In a stepper motor the angular displacement

(A) can be precisely controlled.

(B) it cannot be readily interfaced with micro computer based controller.

(C) the angular displacement cannot be precisely controlled.

(D) it cannot be used for positioning of work tables and tools in NC machines.

Ans: A

124. The power factor of a squirrel cage induction motor is

(A) low at light load only.

(B)   low at heavy load only.

(C) low at light and heavy load both.

(D) low at rated load only.

Ans: A

125. The generation voltage is usually

(A) Between 11 KV and 33 KV.(B) Between 132 KV and 400 KV.

(C) Between 400 KV and 700 KV.   (D) None of the above.

Ans: A

126When a synchronous motor is running at synchronous speed, the damper winding

produces

(A) Damping torque.

(B) Eddy current torque.

(C) Torque aiding the developed torque.

(D) No torque.

Ans: D

127 If a transformer primary is energized from a square wave voltage source, its output

voltage will be

(A) A square wave.(B) A sine wave.

(C) A triangular wave.(D) A pulse wave.

Ans: **A**

128.In a 3 - phase induction motor the maximum torque

(A) is proportional to rotor resistance r2

(B) Does not depend on r_2

(C) None of the above.

Ans: B

129.A 3 phase, 400 V, 50 Hz synchronous motor is operating at zero power factor lagging with respect to the excitation voltage. The armature reaction mmf. produced by the armature current will be

(A) Demagnetizing

(B) Magnetizing

(C) cross-magnetizing

(D) None of the above.

Ans: B

130 In a synchronous motor, the torque angle is

(A) the angle between the rotating stator flux and rotor poles

(B) the angle between magnetizing current and back emf

(C) the angle between the supply voltage and the back emf

(D) none of the above.

Ans: A

131.A 3 phase, 400 V, 50 Hz, 4 pole synchronous motor has a load angle of 10° electrical. The equivalent mechanical degrees will be 35.

(A) 10°

(B)5√2 degrees

(C) 5 degrees

(D) 1 degree.

 Ans: C

 132.A 3 phase, 400 V, 50 Hz synchronous motor has fixed excitation. The load on the motor is doubled. The torque angle, φi will become nearly

(A) φr /2

(B) φr

(C)2 φr

(D) √ 2 φr

 (C) Is proportional to r_2

 (D) None of the above.

Ans: B

129.A 3 phase, 400 V, 50 Hz synchronous motor is operating at zero power factor lagging with respect to the excitation voltage. The armature reaction mmf. produced by the armature current will be

(A) Demagnetizing

(B) Magnetizing

(C) cross-magnetizing

(D) None of the above.

Ans: B

130 In a synchronous motor, the torque angle is

(A) the angle between the rotating stator flux and rotor poles

(B) the angle between magnetizing current and back emf

(C) the angle between the supply voltage and the back emf

(D) none of the above.

Ans: A

131.A 3 phase, 400 V, 50 Hz, 4 pole synchronous motor has a load angle of 10° electrical. The equivalent mechanical degrees will be 35.

(A) 10°

(B)5√2 degrees

(C) 5 degrees

(D) 1 degree.

Ans: C

132. A 3 phase, 400 V, 50 Hz synchronous motor has fixed excitation. The load on the motor is doubled. The torque angle, φi will become nearly

(A) φr /2

(B) φr

(C) 2 φr

(D) √2 φr

Ans :C

133. The hunting in a synchronous motor takes place when

(A) friction in bearings is more

(B) air gap is less

(C) load is variable

(D) load is constant.

Ans:C

134. V curves for a synchronous motor represent relation between

(A) field current and speed

(B) field current and power factor

(C) power factor and speed

(D) armature current and field current.

Ans:D

135. The breakdown. torque of a synchronous motor varies as

(A) 1 /(applied voltage)

(B) 1/(applied voltage)^2

(C) Applied voltage

(D) (Applied voltage)^2.

Ans:C

136. Hunting in a synchronous motor cannot be due to

(A) Variable frequency

(B) Variable load

(C) Variable supply voltage

(D) windage friction.

AnsD

137. When the excitation of an unloaded salient pole synchronous motor suddenly gets disconnected

(A) The motor stops

(B) It runs as a reluctance motor at the same speed

(C) it runs at a reluctance motor at a lower speed.

AnsA

138. Which synchronous motor will be smallest in size ?

(A) 5 HP, 500 rpm

(B) 5 HP, 375 rpm

(C) 10 HP, 500 rpm

(D) 10 HP, 375 rpm.

AnsB

139. A synchronous machine has its field winding on the stator and armature winding on the rotor. Under steady running conditions, the air-gap field

(A) Rotates at synchronous speed with respect to stator

(B) Rotates at synchronous speed with direction of rotation of the rotor

(C) remains stationary with respect to stator

(D) remains stationary with respect to rotor.

AnsC

140.If the field of a synchronous motor is under-excited, the power factor will be

(A) Unity

(B) Lagging

(C) Leading

(D) more than unity.

AnsB

141.The name plate of an induction motor reads 3 phase. 400 V, 50 Hz. 0.8 of lagging, 1440 rpm. On similar lines the name plate of a synchronous motor should read

(A) 3 phase, 400 V, 50 Hz, 0.8 pf lagging, 1500 rpm

(B) 3 phase, 400 V. 50 Hz, 0.8 pf leading, 1500 rpm

(C) 3 phase, 400 V, 50/60 Hz, 0.8 pf lagging, 1500 rpm

(D) 3 phase. 400 V, 50/60 Hz, 0.8 pf leading, 1500 rpm.

142. In which coil the emf generated will be more, for given flux distribution and number of turns

(A) Full pitch coil

(B) Short pitch coil

(C) Long pitch coil

(D) Equal emf will be generated in all cases.

AnsA

143.For a synchronous motor, the ratio starting torque/running torque is

(A) infinite

(B) 1.0

(C) 0.5

(D) 0.

AnsD

144. Synchronous motors for power factor correction operate at

(A) normal load with minimum excitation

(B) normal load with zero excitation

(C) no load and greatly over-excited fields

(D) no load and under-excited fields.

AnsC

145. The construction of a synchronous motor resembles which of the following machines?

(A) Slip ring induction motor

(B) DC shunt generator

(C) Single phase reluctance motor

(D) DC compound motor.

AnsB

146.The construction of a synchronous motor resembles which of the following machine ?

(A) An induction motor

(B) A rotor converter

(C) An alternator

(D) A series motor.

AnsC

147. In a synchronous motor, "hunting" may be due to variation in any of the following EXCEPT:

(A) Load

(B) Supply voltage

(C) Frequency

(D) Winding friction.

AnsD

148. A synchronous motor is switched on to supply with its field winding short-circuited, the motor will?

(A) Not start

(B) Nothing happen.

(C) Start and run as induction motor

(D) Start as induction motor and run as synchronous motor.

AnsD

149. In a synchronous motor, at no load, the armature current is

(A) In phase with the applied voltage

(B) Leading the applied voltage by 90°

(C) Lagging the applied voltage by 90°

(D) Zero.

AnsB

150. In a synchronous motor, during hunting when the rotor speed exceeds the synchronous speed

(A) Field excitation increases

(B) Harmonics are developed

(C) Negative phase sequence currents come into action

(D) Damper bars develop induction generator torque.

AnsD

151. For a synchronous motor when V is the supply voltage, the breakdown torque will be proportional to

(A) V2

(B) V

(C) 1/ V

(D) 1/ V2

AnsA

152. When the field winding of an unloaded salient pole synchronous motor is open - circuited, the motor will

(A) burn with dense smoke

(B) stop

(C) run as induction motor

(D) function as static condenser

AnsB

153. In case one of the three phases of a synchronous motor is short-circuited, the motor will

(A) Not start

(B) Get overheated

(C) Burn out

D) Run normally.

AnsB

154. The fact that a synchronous motor with salient poles will operate, even if field current is reduced to zero, can be explained by

(A) Magnetization of rotor poles by stator magnetic field

(B) Rotating magnetic field of the rotor

(C) Rotating magnetic field of the stator

(D) Interlocking action between stator and rotor fields.

AnsA

155. The negative phase sequence in a three phase synchronous motor exists when the motor is

(A) Under loaded

(B) Overloaded

(C) Supplied with unbalanced voltage

(D) Hot.

AnsC

156. The regulation of a synchronous motor is

(A) 0%

(B) 1%

(C) 50%

(D) 100%.

AnsA

157. In a synchronous motor, the angle between the rotor poles and stator poles is known as

(A) synchronizing angle

(B) torque angle

(C) angle of retardation

(D) power factor angle.

AnsB

158. Single phase motors are commercially manufactured up to

(A) 1H.P.

(B) 2 H.P.

(C) 5 H.P.

(D) 10 H.P.

AnsB

159.The direction of rotation of universal motor can be reversed by

(A) Reversing the supply terminals

(B) Switching over from ac to dc

(C) Interchanging the brush leads

(D) Any of the above.

AnsC

160. A universal motor operates on

(A) Constant speed and varying load

(B) Constant load and varying speed

(C) Approximately constant speed and load

(D) Synchronous speed with varying load.

AnsC

161. Which of the following single phase motors will operate at high power factor ?

(A) Shaded pole motor

(B) Capacitor run motor

(C) Split phase motor

(D) Capacitor start motor.

AnsB

162.A motor suitable for signaling device is

(A) Induction motor

(B) DC shunt motor

(C) DC series motor

(D) Reluctance motors.

AnsD

163. Which capacitor is preferred in case of single phase motor

(A) Paper capacitor

(B) Ceramic capacitor

(C) Mica capacitor

(D) Electrolytic capacitor.

AnsD

164. The motor used for driving the record player deck is

(A) dc series motor

(B) synchronous motor

(C) hysteresis motor

(D) dc shunt motor.

AnsC

165. When a motor speed of 5000 rpm is required, which motor will you select ?

(A) Capacitor start motor

(B) Shaded pole motor

(C) Hysteresis motor

(D) Universal motor.

AnsD

166. As compared to other single phase ac motors a universal motor has

(A) Speed ratio of 2: 1 on ac/dc

(B) Highest efficiency on 50 Hz supply

(C) High horse power/kg ratio

(D) all of the above.

AnsC

167. When a universal motor is operated on no load, its speed is limited by

(A) Armature reaction

(B) Armature weight

(C) windage and friction

(D) Supply voltage frequency.

AnsC

168. Which of the following applications make use of a universal motor ?

(A) Floor polishing machine

(B) Oil expeller

(C) Portable tools

(D) Lathe machines.

AnsC

169. In portable tools the speed of the driven shaft is reduced by

(A) Gearing

(B) Belt drive

(C) Chain drive

(D) Fluid coupling.

AnsA

170. For ceiling fans generally the single phase motor used is

(A) Split phase type

(B) Capacitor start type

(C) Capacitor start and run type

(D) Permanent capacitor type.

AnsD

171. The synchronous speed of a linear induction motor does not depend on

(A) Width of pole pitch

(B) Number of poles

(C) Supply frequency

(D) Any of the above

Ans: B

172. A salient pole synchronous motor is running at no load. Its field current is switched off. The motor will

(A) Come to stop.

(B) Continue to run at synchronous speed.

(C) Continue to run at a speed slightly more than the synchronous speed.

(D) Continue to run at a speed slightly less than the synchronous speed.

Ans: B

173. The emf induced in the primary of a transformer

(A) Is in phase with the flux.

(B) Lags behind the flux by 90 degree.

(C) Leads the flux by 90 degree.

(D) Is in phase opposition to that of flux.

Ans: C

174. The frequency of the rotor current in a 3 phase 50 Hz, 4 pole induction motor at full load speed is about

(A) 50 Hz.

(B) 20 Hz.

(C) 2 Hz.

(D) Zero.

Ans: C

175. The two windings of a transformer is

(A) Conductively linked.

(B) Inductively linked.

(C) Not linked at all.

(D) Electrically linked.

Ans : B

176. The d.c. series motor should always be started with load because

(A) At no load, it will rotate at dangerously high speed.

(B) It will fail to start.

(C) It will not develop high starting torque.

(D) All are true.

Ans: A

177. If a transformer primary is energized from a square wave voltage source, its output voltage will be

(A) A square wave.

(B) A sine wave.

(C) A triangular wave.

(D) A pulse wave.

Ans: A

178. The power factor of a squirrel cage induction motor is

(A) Low at light load only.

(B) Low at heavy load only.

(C) Low at light and heavy load both.

(D) Low at rated load only.

Ans: A

179. In a d.c. machine, the armature mmf is

(A) Stationary w.r.t. armature.

(B) Rotating w.r.t. field.

(C) Stationary w.r.t. field.

(D) Rotating w.r.t. brushes.

Ans: C

180. In a transformer the voltage regulation will be zero when it operates at

(A) Unity p.f.

(B) Leading p.f.

(C) Lagging p.f.

(D) Zero p.f. leading.

Ans: B

181.The primary winding of a 220/6 V, 50 Hz transformer is energised from 110 V, 60 Hz supply. The secondary output voltage will be

(A) 3.6 V.

(B) 2.5 V.

(C) 3.0 V.

(D) 6.0 V.

Ans: C

182. In case of split phase motors the phase shift is usually limited to

(A) 3 degrees

(B) 60 degrees

(C) 90 degrees

(D) 150 degrees.

Ans:A

183. The capacitance of a small single phase motor will be of the order of

(A) Kilo farads

(B) Few hundred farads

(C) Farads

(D) Micro or pico farads.

Ans:D

184. The type of starting relay used on single phase hermetic motor is

(A) hot wire relay

(B) timing relay

(C) current coil relay

(D) voltage coil relay

Ans:C

185. Reluctance motors are

(A) doubly excited

(B) singly excited

(C) either doubly excited or singly excited

(D) none of the above.

Ans:B

186. Electric motors are generally designed to have maximum efficiency at

(A) full load

(B) near full load

(C) half load

(D) near half load.

Ans:B

187.Which of the following is non-reversible motor ?

(A) Universal motor

(B) Capacitor start split phase motor

(C) Resistance start split phase motor

(D) Permanent split capacitor motor.

Ans:C

188. Which motor is generally used for electric shavers ?

(A) Shaded pole motor

(B) Hysteresis motor

(C) Reluctance motor

(D) Universal motor.

Ans:D

189. The motor useful for signaling and timing device is

(A) Reluctance motor

(B) Shaded pole motor

(C) Hysteresis motor

(D) Two value capacitor motor.

Ans:A

190. A motor generally used in toys is

(A) Hysteresis motor

(B) Shaded pole motor

(C) Two value capacitor motor

(D) Reluctance motor.

Ans:B

191.Single phase motors generally get over heated due to

(A) Overloading

(B) Short windings

(C) Bearing troubles

(D) Any of the above.

Ans:D

192. Which of the following is a reversible motor ?

(A) Universal motor

(B) Capacitor start split phase motor

(C) Both (A) and (B) above

(D) None of the above.

Ans:C

193. The starting torque of a single phase induction motor is

(A) uniform

(B) high

(C) low

(D) zero.

Ans:D

194. Which of the following motors will operate at high power factor ?

(A) Universal motor

(B) Capacitor start motor

(C) Capacitor run motor

(D) Split phase motor.

Ans:C

195. Which of the following statements about reluctance motor is not true ?

(A) It is self starting

(B) It runs at constant speed

(C) It needs no dc excitation for its rotor

(D) It can operate on ac as well as dc.

Ans:D

196. The rotor for a hysteresis motor

(A) is made of chrome steel

(B) has high retentivity

(C) has high hysteresis loss

(D) should have all above properties.

Ans:D

197.The speed of a universal motor can be controlled by

(A) introducing a variable resistance in series with the motor

(B) tapping the field at various points

(C) centrifugal mechanisms

(D) any of the above.

Ans:D

198.What could be the smallest size of a universal motor?

(A)1/10 HP

(B)1/20 HP

(C)1/20 HP

(D)1/2000 HP.

Ans:C

199. Single phase induction motor can be made self starting by

(A) adding series combination of a capacitor and auxiliary winding in parallel with the   main winding

(B) adding an auxiliary winding in parallel with the main winding

(C) adding an auxiliary winding in series with a capacitor and the main winding.

Ans:A

200. The following motor is popularly used in driving a refrigerator

(A) d.c. shunt motor

(B) universal motor

(C) plain cage induction motor

(D) squirrel cage induction motor.

Ans:B

201. Motor used in driving a tape recorder is

(A) Hysteresis motor

(B) Synchronous motor

(C) Induction motor

(D) Universal motor.

Ans:A

202.Speed of a repulsion motor at no load is

(A) low

(B) very low

(C) high

(D) dangerously high.

Ans:D

203.Operation of a hysteresis motor can be explained on the basis of

(A) cross field theory

(B) continuously revolving magnetic flux

(C) pulsating magnetic flux

(D) intermittently revolving magnetic flux.

Ans:D

204. The main drawback of a shaded pole motor is

(A) Low efficiency

(B) Low starting torque

(C) Very little overload capacity

(D) All of these.

Ans:D

205. Consider the following single-phase motors :

I. Capacitor start motor

II. Capacitor start and run motor

III. Permanent split capacitor motor

IV. Shaded pole motor.

The correct sequence of the increasing order of their costs is

(A) IV, III, II, I

(B) IV, III, I, II

(C) III, IV, II, I

(D) 111, IV, I, II.

Ans:A

206A universal motor runs at

(A) higher speed with dc supply and with less sparking

(B) higher speed with ac supply and with less sparking

(C) same speed with both ac and dc supplies

(D) higher speed with ac supply but with increased sparking at the brushes.

Ans:C

207 The main reason of using a hysteresis motor , for high quality tape recorders and record players is that

(A) its speed is constant (synchronous)

(B) it develops extremely steady torque

(C) it requires no centrifugal switch

(D) its operation is not affected by mechanical vibrations.

Ans:A

208.A fluctuating voltage supply is detrimental to a refrigerator motor, but not to a ceiling fan motor, although both are single-phase induction motors because, the refrigerator motor

(A) is made more robust than the fan motor

(B) is subjected to short duty cycle but the fan motor is subjected to continuous duty

(C) is enclosed in a sealed unit while the fan motor is open to the environment

(D) load is constant, but the fan motor load is voltage dependent.

Ans:C

209. A capacitor selected for capacitor-run motor should be rated for

(A) peak voltage

(B) rms voltage

(C) Average voltage.

Ans:A

210. If a single phase motor fails to start, the probable cause may be

(A) Open in auxiliary winding

(B) Open in main winding

(C) Blown fuses

(D) Any of the above.

Ans:D

211. The reluctance torque in a motor is present when the reluctance seen by

(A) rotor mmf varies

(B) rotor mmf remains constant

(C) stator mmf is constant.

(D)stator mmf varies

Ans:A

212. Which single phase motor has relatively high power factor ?

(A) Universal motor

(B) Split phase motor

(C) Repulsion motor

(D) Synchronous motor.

Ans:A

213. Which motor would you select for vacuum cleaners ?

(A) Universal motor

(B) Repulsion motor

(C) Hysteresis motor

(D) Reluctance motor.

Ans:A

214.If the ceiling fan, when switched on, runs at slow speed in the reverse direction, it can be concluded that

(A) winding has burnt out

(B) bearings are worn out

(C) capacitor is ineffective

(D) none of the above.

Ans:C

215. For how many poles is a split-phase motor wound if it operates at 1750 rpm at full load from a 60 Hz source ?

(A) 2 poles

(B) 4 poles

(C) 6 poles

(D) 12 poles.

Ans:B

216. Which of the following capacitor-start split phase motor will have the largest value of capacitance ?

(A)1/8 HP, 3450 rpm

(B) 1/4 HP, 1725 rpm

(C) 1/2 HP, 1140 rpm

(D) 3/4 HP, 1140 rpm.

Ans:D

217.The speed of a split phase induction motor can be reversed by reversing the leads of

(A) auxiliary winding

(B) main winding

(C) either of (A) and (B) above

(D) speed cannot be reversed.

Ans:C

218 In a 3-phase synchronous motor

(A) the speed of stator MMF is always more than that of rotor MMF.

(B) the speed of stator MMF is always less than that of rotor MMF.

(C) the speed of stator MMF is synchronous speed while that of rotor MMF is zero.

(D) rotor and stator MMF are stationary with respect to each other.

Ans: D

Because, Motor is magnetically locked into position with stator, the rotor poles are engaged with stator poles and both run synchronously in same direction Therefore, rotor & stator mmf are stationary w.r.t each other.

219In a capacitor start single-phase induction motor, the capacitor is connected

(A) in series with main winding.

(B) in series with auxiliary winding.

(C) in series with both the windings.

(D) in parallel with auxiliary winding.

Ans: B

To make single phase motor self start. We split the phases at 90 degree. Hence, motor behaves like a two phase motor.

220A synchro has

(A) a 3-phase winding on rotor and a single-phase winding on stator.

(B) a 3-phase winding on stator and a commutator winding on rotor.

(C) a 3-phase winding on stator and a single-phase winding on rotor.

(D) a single-phase winding on stator and a commutator winding on rotor.

Ans: C

Synchros : The basic synchro unit called a synchro transmitter. It's construction similar to that of a Three phase alternator.

221As the voltage of transmission increases, the volume of conductor

(A) increases.(B) does not change.

(C) decreases.(D) increases proportionately.

Ans: C

Decreases due to skin effect.

222.The size of the feeder is determined primarily by

(A) the current it is required to carry.

(B) the percent variation of voltage in the feeder.

(C) the voltage across the feeder.

(D) the distance of transmission.

Ans: A

Size of conductor depends upon amount of current flow.

223The boundary of the protective zone is determined by the

(A) Location of CT(B) sensitivity of relay used

(C) Location of PT(D) None of these

Ans: B

The boundary of the protective zone is determined by the sensitivity of relay used. If the relay is more sensitive, the protective zone will be increased.

224 In a three phase transformer, if the primary side is connected in star and secondary side is connected in delta, what is the angle difference between phase voltage in the

two cases.

(A) delta side lags by -300.(B) star side lags by -300.

(C) delta side leads by 300.(D) star side leads by -300.

Ans: C

This is vector group and has +300 displacement. Therefore, delta side leads by +300.

225 To achieve low PT error, the burden value should be

_____.

(A) low   (B) high

(C) medium   (D) none of the above

Ans: A

In a Potential transformer, burden should be in permissible range to maintain errorless measurement.

226Slip of the induction machine is 0.02 and the stator supply frequency is 50 Hz. What will be the frequency of the rotor induced emf?

(A)10 Hz.   (B) 50 Hz.

(C) 1 Hz.   (D) 2500 Hz

Ans : B

227 A capacitor start type single phase induction motor has its capacitor replaced by an inductor of equivalent reluctance value. If the motor is now switched on to the supply, it will

(A) Run in the reverse direction

(B) Not run

(C) Run with more noise and low torque

(D) Start but not take the load.

AnsB

228 In a transformer, zero voltage regulation at full load is [GATE 2007]

(A) Not possible

(B) Possible at unity Power factor load

(C) Possible at leading Power factor load

(D) Possible at lagging Power factor load

Ans:C

229 The DC motor, which can provide zero speed regulation at full load without any controller is [GATE 2007]

(A) Series

(B) Shunt

(C) Cumulative Compound

(D) Differential Compound

Ans:B

230 In 8 - pole wave connected motor armature, the number of parallel paths are

(A) 8

(B) 4

(C) 2

(D) 1

Ans: C

231 In a transformer the voltage regulation will be zero when it operates at

(A) Unity p.f.

(B) Leading p.f.

(C) Lagging p.f.

(D) Zero p.f. leading.

Ans: B

232 In a stepper motor the angular displacement

(A) Can be precisely controlled.

(B) It cannot be readily interfaced with micro computer based controller.

(C) The angular displacement cannot be precisely controlled.

(D) It cannot be used for positioning of work tables and tools in NC machines.

Ans: A

233 The power factor of a squirrel cage induction motor is

(A) Low at light load only.

(B) Low at heavy load only.

(C) Low at light and heavy load both.

(D) Low at rated load only.

Ans: A

234 The generation voltage in India is usually

(A) Between 11 KV and 33 KV.

(B) Between 132 KV and 400 KV.

(C) Between 400 KV and 700 KV.

(D) None of the above.

Ans: A

235 When a synchronous motor is running at synchronous speed, the damper winding produces

(A) Damping torque.

(B) Eddy current torque.

(C) Torque aiding the developed torque.

(D) No torque.

Ans: D

236 If a transformer primary is energized from a square wave voltage source, its output voltage will be

(A) A square wave.

(B) A sine wave.

(C) A triangular wave.

(D) A pulse wave.

Ans: A

237 A salient pole synchronous motor is running at no load. Its field current is switched off. The motor will

(A) Come to stop.

(B) Continue to run at synchronous speed.

(C) Continue to run at a speed slightly more than the synchronous speed.

(D) Continue to run at a speed slightly less than the synchronous speed.

Ans: B

238 The frequency of the rotor current in a 3 phase 50 Hz, 4 pole induction motor at full load speed is about

(A) 50 Hz.

(B) 20 Hz.

(C) 2 Hz.

(D) Zero.

Ans: C

239 The speed of a dc motor can be controlled by varying

(A) Its flux per pole

(B) Resistance of armature circuit

(C) Applied voltage

(D) All of the above

Ans: D

240 Regarding Ward-Leonard system of speed control which statement is false?

(A) It is usually used where wide and very sensitive speed control is required

(B) It is used for motors having ratings from 750kW to 4000kW

(C) Capital outlay involved in the system is right since it uses two extra machines

(D) It gives a speed range of 10:1 but in one direction only

Ans: D

241 A stepper motor is

(A) A dc motor.

(B) A single-phase ac motor.

(C) A multi-phase motor.

(D) A two phase motor.

Ans: D

Stepper motor works on 1-phase-ON or 2-phase –ON modes of operation

242. A motor which can conveniently be operated at lagging as well as leading power factors is the

(A) Squirrel cage induction motor.

(B) Wound rotor induction motor.

(C) Synchronous motor.

(D) DC shunt motor.

Ans: C

243 The d.c. series motor should always be started with load because

(A) At no load, it will rotate at dangerously high speed.

(B) It will fail to start.

(C) It will not develop high starting torque.

(D) All are true.

Ans: A

244 The 'sheath' is used in cable to

(A) Provide strength to the cable.

(B) Provide proper insulation.

(C) Prevent the moisture from entering the cable.

(D) Avoid chances of rust on strands.

Ans: A

The sheath in underground cable is provided to give mechanical strength.

245 The drive motor used in a mixer-grinder is a

(A) DC motor.

(B) Induction motor.

(C) Synchronous motor.

(D) Universal motor.

Ans: D

The universal motor is suitable for AC & DC both supply systems.

246 A balanced three-phase, 50 Hz voltage is applied to a 3 phase, 4 pole, induction motor. When the motor is delivering rated output, the slip is found to be 0.05. The speed of the rotor m.m.f. relative to the rotor structure is

(A) 1500 r.p.m.

(B) 1425 r.p.m.

(C) 25 r.p.m.

(D) 75 r.p.m.

Ans: D

NS = 120f /P = 120 x 50 /4 =1500rpm

N = NS (1-s) = 1500 (1-0.05) = 1425

Relative speed = 1500 – 1425 = 75 rpm

247 The primary winding of a 220/6 V, 50 Hz transformer is energised from 110 V, 60 Hz supply. The secondary output voltage will be

(A) 3.6 V.

(B) 2.5 V.

(C) 3.0 V.

(D) 6.0 V.

Ans: C

248 The current from the stator of an alternator is taken out to the external load circuit through

(A) Slip rings.

(B) Commutator segments.

(C) Solid connections.

(D) Carbon brushes.

Ans: C

249 In a 3 – phase induction motor the maximum torque

(A) Is proportional to rotor resistance r2 .

(B) Does not depend on r2 .

(C) Is proportional to square root of r2 .

(D) Is proportional to square of r2 .

Ans: B

250 In a d.c. machine, the armature mmf is

(A) Stationary w.r.t. armature.

(B) Rotating w.r.t. field.

(C) Stationary w.r.t. field.

(D) Rotating w.r.t. brushes.

Ans: C

251.The emf induced in the primary of a transformer

(A) Is in phase with the flux.

(B) Lags behind the flux by 90 degree.

(C) Leads the flux by 90 degree.

(D) Is in phase opposition to that of flux.

Ans: C

252. The relative speed between the magnetic fields of stator and rotor under steady state operation is zero for a

(A) DC machine.

(B) 3 phase induction machine.

(C) Synchronous machine.

(D) Single phase induction machine.

Ans: all options are correct

253. As the voltage of transmission increases, the volume of conductor

(A) Increases.

(B) Does not change.

(C) Decreases.

(D) Increases proportionately.

Ans: C

Due to skin effect

254. In a 3-phase synchronous motor

(A) The speed of stator MMF is always more than that of rotor MMF.

(B) The speed of stator MMF is always less than that of rotor MMF.

(C) The speed of stator MMF is synchronous speed while that of rotor MMF is zero.

(D) Rotor and stator MMF are stationary with respect to each other.

Ans: D

Motor is magnetically locked into position with stator, the rotor poles are engaged with stator poles and both run synchronously in same direction.

256 An alternator is delivering rated current at rated voltage and 0.8 power-factor lagging case. If it is required to deliver rated current at

rated voltage and 0.8 power-factor leading, the required excitation will be

(A) Less.

(B) More.

(C) More or less.

(D) The same.

Ans: B

Over excitation gives leading power factor and under excitation gives lagging p.f .

257. Out of the following methods of heating the one which is independent of supply frequency   is

(A) Electric arc heating

(B) Induction heating

(C) Electric resistance heating

(D) Dielectric heating

Ans: C

258. In a capacitor start single-phase induction motor, the capacitor is connected

(A) In series with main winding.

(B) In series with auxiliary winding.

(C) In series with both the windings.

(D) In parallel with auxiliary winding.

Ans: B

To make single phase motor self start. We split the phases at 90 degree. Hence, motor behaves like a two phase motor..

259. A ceiling fan uses

(A) Split-phase motor.

(B) Capacitor start and capacitor run motor.

(C) Universal motor.

(D) Capacitor start motor.

Ans: D

To give starting torque and to maintain speed.

260 .The torque-speed characteristics of an a.c. operated universal motor has a _____ characteristic and it_____ be started under no-load condition.

(A) Inverse, can

(B) Nearly inverse, can

(C) Inverse, cannot

(D) Nearly inverse, cannot

Ans: C

N direct proportional to 1/ T

261. In the heating process of the _____type a simple method of temperature control is possible by means of a special alloy which loses its magnetic properties at a particular high temperature and regains them when cooled to a temperature below this value.

(A) Indirect induction over

(B) Core type induction furnace

(C) Coreless induction furnace

(D) High frequency eddy current

Ans: D

Magnetic property of alloy changes with change of the temperature and Heat is produced due to eddy current = (i)2 * R and i proportional to (f)2

262. In order to reduce the harmful effects of harmonics on the A.C. side of a high voltage D.C. transmission system _____are provided.

(A) Synchronous condensers

(B) Shunt capacitors

(C) Shunt filters

(D) Static compensators

Ans: C

$X_c = 1/\omega C$

263.An a.c. tachometer is just a _____with one phase excited from the carrier frequency.

(A) Two-phase A.C. servomotor

(B) Two-phase induction motor

(C) A.C. operated universal motor

(D) Hybrid stepper motor.

Ans: D

It is a special purpose machine. It's stator coil can be energized by electronically switched current.

264. The rotor frequency for a 3 phase 1000 RPM 6 pole induction motor with a slip of 0.04 is_____Hz

(A) 8

(B) 4

(C) 6

(D) 2

Ans: D

f = N P/ 120 = 1000* 6/120 = 50 Hz

Rotor frequency fr=s * f = 0.04 * 50= 2.0 Hz

265. The speed-torque characteristics of a DC series motor are approximately similar to those of the _____ motor.

(A) Universal

(B) Synchronous

(C) DC shunt

(D) Two-phase

Ans: A

Universal motor has same characteristics as DC series motor

It is known as an a.c series motor.

266.In case of a universal motor, torque pulsation is minimized by _____.

(A) Load inertia

(B) Rotor inertia

(C) Both rotor and load inertia

(D) None of the above

Ans: C

267. A hysteresis motor

(A) Is not a self-starting motor.

(B) Is a constant speed motor.

(C) Needs dc excitation.

(D) Cannot be run in reverse speed.

Ans: B

268. The most suitable servomotor for low power applications is

(A) A dc series motor.

(B) A dc shunt motor.

(C) An ac two-phase induction motor.

(D) An ac series motor.

Ans: B

269. The size of a conductor used in power cables depends on the

(A) Operating voltage.

(B) Power factor.

(C) Current to be carried.

(D) Type of insulation used.

Ans: C

270.In a DC motor, unidirectional torque is produced with the help of

(A) Brushes

(B) Commutator

(C) End-plates

(D) Both a and b

Ans: D

271 The counter emf of dc motor

(A) Often exceeds the supply voltage

(B) Aids the applied voltage

(C) Helps in energy conversion

(D) Regulates its armature voltage

Ans: C

272.The E_b/V ratio of a dc motor is an indication of its

(A) Efficiency

(B) Speed regulation

(C) Starting torque

(D) Running Torque

Ans: A

273. The induced emf in the armature conductors of a dc motor is

(A) Sinusoidal

(B) Trapezoidal

(C) Rectangular

(D) Alternating

Ans: A

274. A dc motor can be looked upon as dc generator with the power flow

(A) Reduced

(B) Reversed

(C) Increased

(D) Modified

Ans:B

275. A series motor is best suited for driving

(A) Lathes

(B) Cranes and hoists

(C) Shears and punches

(D) Machine tools

Ans: B

276.The Ta/Ia graph of a dc series motor is a

(A) Parabola from no load to overload

(B) Straight line throughout

(C) Parabola throughout

(D) Parabola upto full load and a straight line at overloads

Ans: D

277. When load is removed, ...motor will run at the highest speed.

(A) Shunt

(B) Cumulative - compound

(C) Differential compound

(D) Series

Ans: D

278. The power factor of a squirrel cage induction motor is

(A) Low at light loads only

(B) Low at heavy loads only

(C) Low at light and heavy loads both

(D) Low at rated load only

Ans: A

279.In a d.c. machine, the armature mmf is

(A) stationary w.r.t. armature.(B) rotating w.r.t. field.

(C) stationary w.r.t. field.(D) rotating w.r.t. brushes

Ans: C

280.The maximum power in cylindrical and salient pole machines is obtained respectively

at load angles of

(A)90degree ,90 degree   (B)<90 degree,90 degree

(C)90 degree,>90 degree   (D)90 degree, <90 degree

Ans: D

281.The primary winding of a 220/6 V, 50 Hz transformer is energized from 110 V, 60 Hz

supply. The secondary output voltage will be

(A) 3.5 V.   (B) 2.5 V.

(C) 3.0 V.   (D) 7.0 V.

Ans: C

282.The emf induced in the primary of a transformer

(A) is in phase with the flux.(B) lags behind the flux by 90 degree.

(C) is in phase opposition to that of flux. (D) leads the flux by 90 degree.

Ans: D

283.The relative speed between the magnetic fields of stator and rotor under steady state

operation is zero for a

(A) dc machine.(B) 3 phase induction machine.

(C) synchronous machine.(D) single phase induction machine.

Ans: all options are correct

284.The current from the stator of an alternator is taken out to the external load circuit through

(A) slip rings.(B) commutator segments.

(C) solid connections.(D) carbon brushes.

Ans: C

285.A motor which can conveniently be operated at lagging as well as leading power factors is the

(A) squirrel cage induction motor. (B)   synchronous motor

.(C) DC shunt motor.   (D) wound rotor induction motor

Ans: B

286 In a dc shunt motor the terminal voltage is halved while the torque is kept constant ω and armature current $'I_a'$ will be

(A) Both ω and I_a are doubled.

(B) ω is constant and I_a is doubled.

(C) ω is doubled while I_a is halved.

(D) ω is constant but I_a is halve

Ans: B

$N \alpha V - IaRa \qquad$ or $N \alpha Eb$

$T \alpha Ia \phi , \phi \alpha Ia$

$T \alpha I^2_a$

287 A balanced three-phase, 50 Hz voltage is applied to a 3 phase, 4 pole, induction motor. When the motor is delivering rated output, the slip is found to be 0.05. The speed of the rotor m.m.f. relative to the rotor structure is

(A)1500 r.p.m.(B)   1425 r.p.m. (C) 25 r.p.m.   (D)   75 r.p.m.

Ans: D

NS = 120f /P = 120 x 50 /4   =1500rpm

N = NS (1-s) = 1500 (1-0.05) = 1425

So relative speed   = 1500 - 1425 = 75 rpm

288 An alternator is delivering rated current at rated voltage and 0.8 power-factor lagging case. If it is required to deliver rated current at rated voltage and 0.8 power-factor leading, the required excitation will be

(A) less.   (B) more.

(C) more or less.(D) the same.

Ans: B

Over excitation gives leading power factor and under excitation gives lagging p.f .

Q289A 1:5 step-up transformer has 120V across the primary and 600 ohms resistance across the secondary. Assuming 100% efficiency, the primary current equals

(A) 0.2 Amp.   (B) 5 Amps.

(C) 10 Amps.   (D) 20 Amps.

Ans: A

$I_1 = V_1 / R_1 = 120/600 = 0.2$ ($\eta = 100\%$, losses are zero . $V_1 = VR = I_1 R_1$)

290A dc shunt generator has a speed of 800 rpm when delivering 20 A to the load at the terminal voltage of 220V.   If the same machine is run as a motor it takes a line current of 20A from 220V supply.   The speed of the machine as a motor will be

(A) 800 rpm.   (B) more than 800 rpm.

(C) less than 800 rpm. (D) both higher or lower than 800 rpm.

Ans: C

$N_g = E_g (60A / \phi pz)$, $E_g = V + I_a R_a$; in generator

$Nm = E_b (60A / \phi pz)$, $E_b = V - I_a R_a$; in motor

$E_g > E_b$ for same terminal voltage

Therefore, $Ng > N\,m$

291.A 50 Hz, 3-phase induction motor has a full load speed of 1440 r.p.m. The number of poles of the motor are

(A) 4.   (B) 6.

(C) 12 (D) 8.

Ans: A

292 A 4 pole lap wound dc shunt motor rotates at the speed of 1500 rpm, has a flux of 0.4 mWb and the total number of conductors are 1000. What is the value of emf?

(A) 100 Volts. **(B)** 0.1 Volts.

(C) 1 Volts   **(D)** 10 Volts.

Ans: D

Given $N = 1500$ rpm, $\Phi = 0.4$ mWb, $Z = 1000$, $P = 4$, & $A = 4$

Therefore, $E_b = N\Phi PZ / 60\,A$

$= 1500 \times 0.4 \times 4 \times 1000 \times 10^{-3} / 60 \times 4 = 60/6 = 10$ volts.

Q293 The synchronous reactance of the synchronous machine is

_____.

(A) Ratio between open circuit voltage and short circuit current at constant field current

(B) Ratio between short circuit voltage and open circuit current at constant field current

(C) Ratio between open circuit voltage and short circuit current at different field current

(D) Ratio between short circuit voltage and open circuit current at different field current

Ans. A

The Synchronous reactance of a synchronous machine is a total steady state reactance, presented to applied voltage, when rotor is running synchronously without excitation.

Therefore , $X_S = E_f / I_S$

$\qquad\qquad$ = Emf of OC for same I_f / short circuit current

Q294 A 3 stack stepper motor with 12 numbers of rotor teeth has a steep angle of _____.

(A) $12°$ $\qquad\qquad$ **(B)** $8°$

(C) 24°   (D) 10°

Ans. D

Given m = 3, N_r = 12

Step angle = 360 / m x N_r = 360 /3 x 12 = 10°

295 In case of a universal motor, torque pulsation is minimized by

_____.

(A) load inertia   (B) rotor inertia

(C) both rotor and load inertia   (D) none of the above

Ans: C

In a universal motor, torque pulsation is minimized by rotor and load inertia.

296 Oil-filled cable has a working stress of _____ kV/mm

(A) 10   (B) 12

(C) 13   (D) 15

Ans: D

This is defined by dielectric strength of mineral oil i.e. 15 kV/mm.

Q297 Inverse definite minimum time lag relay is also called

_____

(A) pilot relay.   (B) differential relay.

(C) over current relay.(D) directional over current relay.

Ans: B

Inverse definite minimum time lag relay characteristic is inverse but minimum time is fixed. The operating time is inversely proportional to the magnitude of actuating quantity.

298 Specific heat of nickel -chrome is _____

(A) 0.112   (B) 0.106.

(C) 0.108.   (D) 0.110.

Ans: None of these

Specific heat of Nickel-Chrome is 440 J/kg°C to 450 J/kg°C

299 The speed-torque characteristics of a DC series motor are approximately similar to those of the _____motor.

(A) universal (B) synchronous

(C) DC shunt (D) two-phase

Ans: A

Universal motor has same characteristics as DC series motor and also known as an a.c series motor.

300 The rotor frequency for a 3 phase 1000 RPM 6 pole induction motor with a slip of

0.04 is_____Hz

(A)8   (B) 4

(C)6   (D) 2

Ans: D

Given: N=1000 rpm ; P= 6; s= 0.04;

and f= N P/ 120

= 1000 x 6/120

= 50 Hz

Rotor frequency $f_r = s \times f = 0.04 \times 50$

= 2.0 Hz

301 The torque-speed characteristics of an a.c. operated universal motor has _____characteristic and it_____ be started under no-load condition.

(A)inverse, can   (B) nearly inverse, can

(C)inverse, cannot   (D) nearly inverse, cannot

Ans: C

If torque is zero then speed may exceed up to infinite, that is dangerous for machine and machine can be damaged.

N α 1/ T

302 In the heating process of the _____type a simple method of temperature control is possible by means of a special alloy which loses its magnetic properties at a particular high temperature and regains them when cooled to a temperature below this value.

(A) Indirect induction over   (B) core type induction furnace

(C) coreless induction furnace (D) high frequency eddy current

Ans: D

Magnetic property of alloy changes with change of the temperature and Heat is produced due to eddy current = i^2R and i α f^2

303 In order to reduce the harmful effects of harmonics on the A.C. side of a high voltage D.C. transmission system _____are provided.

(A) synchronous condensers   (B) shunt capacitors

(C)shunt filters   (D) static compensators

Ans: C

$X_c = 1/\omega c$

304 An a.c. tachometer is just a_____with one phase excited from the carrier frequency.

(A) two-phase A.C. servomotor   (B) two-phase induction motor

(C) A.C. operated universal motor   (D) hybrid stepper motor.

Ans: D T his is a special purpose machine whose stator coil can be energized by electronically switched current.

305 The torque, in a _____is proportional to the square of the armature current

(A) DC shunt motor   (B) stepper motor

(C) 2-phase servomotor   (D) DC series motor

Ans: D

Ta α Φ.Ia   and   Φ α Ia ; therefore Ta α Ia²

306 The synchronous speed for a 3 phase 6-pole induction motor is 1200 rpm. If the number of poles is now reduced to 4 with the frequency remaining constant, the rotor speed with a slip of 5% will be _____.

(A) 1690 rpm   (B) 1750 rpm

(C) 1500 rpm   (D) 1710 rpm

Ans: D

307 The eddy current loss in an a-c electric motor is 100 watts at 50 Hz. Its loss at 100 Hz will be

(A) 25 watts   **(B)** 59 watts

(C) 100 watts   **(D)** 400 watts

Ans: D

Eddy current losses $\alpha \; f^2$

New loss $\alpha \; (2f)^2$

New loss $\alpha \; 4f^2$

$\therefore$ 4 times

308 The maximum power for a given excitation in a synchronous motor is developed when the power angle is equal to

(A) 0° **(B)** 45°

(C) 60° **(D)** 90°

Ans: A

$P = VI \cos\Phi$, $Pmax = VI$,

$\therefore \Phi = 0^0$

309 A commutator in a d.c. machine

(A) Reduces power loss in armature.

(B) Reduces power loss in field circuit.

(C) Converts the induced a.c armature voltage into direct voltage.

(D) Is not necessary.

Ans: C

As name suggests, it commutes ac into dc.

310. The speed of a d.c. shunt motor at no-load is

(A) 5 to 10%   **(B)** 15 to 20%

(C) 25 to 30%   **(D)** 35 to 40%

higher than its speed at rated load.

Ans: A

Ta α Φ Ia ,, Φ = constant,

$\therefore T \alpha I_a$

$N \alpha E_b / \Phi$ or $N \alpha E_b$ initially E_b less , so speed is less.

311. The efficiency of a transformer is mainly dependent on

(A) core losses.   **(B)** copper losses.

(C) stray losses.   **(D)** dielectric losses.

Ans: A

Core loss has prominent value over other losses

312. When two transformers are operating in parallel, they will share the load as under:

(A) proportional to their impedances.

(B) inversely proportional to their impedances.

(C) 50% - 50%

(D) 25%-75%

Ans: A

High rating transformer has higher impedance.

kVA rating α Impedance of transformer

313. A 3-phase, 400 volts, 50 Hz, 100 KW, 4 pole squirrel cage induction motor with a rated slip of 2% will have a rotor speed of

(A) 1500 rpm   (B) 1470 rpm

(C) 1530 rpm   (D) 1570 rpm

Ans: B

$N = N_S (1-S)$ and $N_S = 120 \, f / p$

$= 120 \times 50 / 4 = 1500$ rpm

$\therefore N = 1500 (1-0.02) = 1470$ rpm

314. The voltage at the two ends of a transmission line are 132 KV and its reactance is 40 ohm. The Capacity of the line is

(A) 435.6 MW   (B) 217.8 MW

(C) 251.5 MW   (D) 500 MW

Ans: A

Line capacity is determined by power of line

$P = (V^2/R)$ or (V^2/Z) when $\cos \Phi = 1$

315. A 220/440 V, 50 Hz, 5 KVA, single phase transformer operates on 220V, 40Hz supply with secondary winding open circuited. Then

(A) Both eddy current and hysteresis losses decreases.

(B) Both eddy current and hysteresis losses increases.

(C) Eddy current loss remains the same but hysteresis loss increases.

(D) Eddy current loss increases but hysteresis loss remains the same.

Ans: A

$P_h = k_h f B^{1.6}{}_m$ and $P_e = k_e f^2 B^2{}_m$.

Therefore, hysteresis and eddy current losses will be decreased when frequency decreases.

316. A synchronous motor is operating on no-load at unity power factor. If the field current is increased, power factor will become

(A) Leading & current will decrease

(B) Lagging & current will increase.

(C) Lagging & current will decrease.

(D) Leading & current will increase.

Ans: A

initially synchronous motor is operating at no load and unity power factor. When field current increases, the excitation will increase. Therefore, p.f will be leading and current will be I CosΦ < I

317. A d.c. shunt motor runs at no load speed of 1140 r.p.m. At full load, armature reaction weakens the main flux by 5% whereas the armature circuit voltage drops by 10%. The motor full load speed in r.p.m. is

(A) 1080   **(B)**   1203

(C) 1000   **(D)**   1200

Ans: A

$N_2 / N_1 = E_{b2}/E_{b1} \times \Phi_1 / \Phi_2$;$\Phi_2 = 0.95\Phi_1$; $E_{b2} = 0.9E_{b1}$

$\therefore N_2 / 1140 = 0.9 \times 1/0.95$

$N_2 = 1080$

318. The introduction of interpoles in between the main pole improves the performance of d.c. machines, because

(A) The interpole produces additional flux to augment the developed torque.

(B) The flux waveform is improved with reduction in harmonics

(C) The inequality of air flux on the top and bottom halves of armature is removed.

(D) A counter e.m.f. is induced in the coil undergoing commutation.

Ans: D

Counter e.m.f is produced, it neutralizes the reactive emf.

319. The rotor power output of a 3-phase induction motor is 15 KW and corresponding slip is 4%. The rotor copper loss will be

(A) 600 W   **(B)** 625 W

(C) 650 W   **(D)** 700 W

Ans: B

Rotor copper losses = rotor input- rotor output

and output = (1-s) input

∴ Input = output/(1-s) = 15000 /1-0.04 = 15625

∴ loss = 15625 -1500 = 625 watt.

320. The direction of rotation of hysteresis motor is reversed by

(A) Shift shaded pole with respect to main pole

(B) Reversing supply lead

(C) Either A or B

(D) Neither A nor B

Ans: A

This motor used single phase, 50Hz supply and stator has two windings. These are connected continuously from starting to running.

321. Low head plants generally use

(A) Pelton Turbines   **(B)** Francis Turbine

(C) Pelton or Francis Turbine **(D)** Kaplan Turbines

Ans: A

In the hysterisis motor, the direction of rotation can be reversed by shifting the shaded pole region with respect to main pole. But not by changing supply lead because it has ac supply.

322. In DC generators, armature reaction is produced actually by

(A) Its field current.   **(B)** Armature conductors.

(C) Field pole winding.   **(D)** Load current in armature.

Ans: D

Because load current in armature gives rise to armature mmf which react with main field mmf.

323Two transformers operating in parallel will share the load depending upon their

(A) Rating.   **(B)** Leakage reactance.

(C) Efficiency.**(D)** Per-unit impedance.

Ans: A

Transformers having higher kVA rating will share more load.

324 As compared to shunt and compound DC motors, the series DC motor will have the highest torque because of its comparatively _____ at the start.

(A) Lower armature resistance. **(B)** Stronger series field.

(C) Fewer series turns. **(D)** Larger armature current.

Ans: D

$T \alpha \Phi I_a$ (before saturation)

$\Phi \alpha I_a$

$T \alpha I^2{}_a$

325 A 400kW, 3-phase, 440V, 50Hz induction motor has a speed of 950 r.p.m. on full-load. The machine has 6 poles. The slip of the machine will be _____.

(A) 0.06 **(B)** 0.10

(C) 0.04 **(D)** 0.05

Ans: D

$N = N_s (1-S)$

$950 = 120 \times 50 (1-S)/6$

$S = 0.05$

326. Reduction in the capacitance of a capacitor-start motor, results in reduced

(A) Noise. **(B)** Speed.

(C) Starting torque.**(D)** Armature reaction.

Ans: C

Reduction in the capacitance reduces starting voltage, which results in reduced starting torque.

327 Regenerative braking

(A) Can be used for stopping a motor.

(B) Cannot be easily applied to DC series motors.

(C) Can be easily applied to DC shunt motors

(D) Cannot be used when motor load has overhauling characteristics.

Ans: B

Because reversal of I_a would also mean reversal of field and hence of E_b

328 At present level of technology, which of the following method of generating electric power from sea is most advantageous?

(A) Tidal power.**(B)** Ocean thermal energy conversion

(C) Ocean currents. **(D)** Wave power.

Ans: A

At present level of technology, tidal power for generating electric power from sea is most advantageous because of constant availability of tidal power.

329. If the field circuits of an unloaded salient pole synchronous motor gets suddenly open circuited, then

(A) The motor stops.

(B) It continues to run at the same speed.

(C) Its runs at the slower speed.

(D) It runs at a very high speed.

Ans: B

The motor continues to run at the same speed because synchronous motor speed does not depend upon load, Nα f.

330 In a salient pole synchronous machine (usual symbols are used):

(A)$xq > xd$ (B)$xq = xd$

(C)$xq < xd$ (D)$xq = 0$

Ans: C

Since reluctance on the q axis is higher, owing to the larger air gap, hence $x_q < x_d$